ADVANCES IN BIOCHEMICAL ENGINEERING

Volume 8

Editors: T. K. Ghose, A. Fiechter, N. Blakebrough

Managing Editor: A. Fiechter

With 95 Figures

Springer-Verlag Berlin
Heidelberg GmbH 1978

ISBN 978-3-662-15464-9   ISBN 978-3-540-35958-6 (eBook)
DOI 10.1007/978-3-540-35958-6

© by Springer-Verlag Berlin Heidelberg 1978
Library of Congress Catalog Card Number 72-152360
Originally published by Springer-Verlag Berlin Heidelberg New York in 1978.
Softcover reprint of the hardcover 1st edition 1978

2152/3140-543210

Contents

Technical Aspects of the Rheological Properties of Microbial Cultures

M. Charles
Biochemical Engineering Group, Department of Chemical Engineering,
Lehigh University, Bethlehem, PA 18015, U.S.A.

Contents

Summary

The rheological properties of culture fluids have a profound effect on the course of a fermentation, the response and reliability of sensors, and on the difficulty of recovery processes. In addition, the rheological properties can be sensitive indicators of the state of a fermentation and can be useful for purposes of control and monitoring. In this paper the fundamentals and important nomenclature of rheology are introduced, the merits of various experimental methods for measuring rheological properties are discussed, critical evaluations are presented of past studies of the rheological properties of culture fluids and the effects of these properties, and suggestions are made for further investigation and development.

1. Introduction

The rheological characteristics of a culture fluid affect directly bulk mixing behavior, all forms of mass transfer, and heat transfer and therefore can have a profound influence not only on the course and outcome of a microbial reaction but also on the response of sensors used for monitoring and control. In addition, the rheological properties of the final culture fluid play a large part in determining the ease or difficulty of recovery and purification operations. Accurate measurement and meaningful correlation of culture fluid rheological properties with the various transport phenomena are of critical importance in the

(1) interpretation and control of basic experiments,
(2) development of rational scale-up procedures,
(3) design and utilization of monitoring and control systems,
(4) interpretation of sensor responses,
(5) control of plant-scale bioprocesses,
(6) development of methods to increase yield and/or productivity,
(7) design and operation of recovery equipment.

It is also important to note that in most cases where rheological properties are controlling factors they are also sensitive indicators of the state of the process and should be considered for purposes of monitoring and control.

Usually, pre-inoculation media exhibit simple water-like rheological behaviour but as the fermentation proceeds the rheological characteristics tend to become more complex and usually more adverse. These changes are caused by increased cell mass and/or the accumulation of appreciable quantities of high molecular weight products such as extracellular polysaccharides. Effects of cell mass are usually most pronounced for mycelial cultures in which the mycelia tend to "structure" the fluid. Thus, a given fungus will cause more troublesome rheological conditions when growing in a mycelial habit than when growing in pellet form. Most bacteria and yeasts at concentrations normally encountered do not have very great effects on rheological conditions.

Insofar as extracellular products are concerned, only polysaccharides exert a strong effect on rheological properties. As opposed to mycelial broths, the rheological properties of polysaccharide-containing culture fluids are determined by the nature of the continuous phase. This difference is significant as will be seen later. Other macromolecules such as proteins do influence rheological properties, particularly in the presence of finely dispersed gas bubbles [1], but such rheological effects are usually negligible in bioprocesses.

Finally, there are cases in which the initial medium itself exhibits complex rheological behavior which is generally associated with moderate to high concentrations of starch, etc. In such cases the hydrolytic action of extracellular enzymes usually decreases the average molecular weight of these nutrients thereby decreasing viscosity quite rapidly during the early reaction stages of the fermentation. Subsequent changes are governed by factors already noted.

Few careful and systematic studies of the rheological properties of culture fluids have been reported and little has been done to explore carefully and to correlate the effects of rheological behavior on reactor performance. Part of the reason for this is the complex nature of the problem which requires simultaneous consideration of rheological, microbiological, mixing and mass and heat transfer effects. In most cases only the microbiological aspect is considered and this frequently results in misinterpretation or in conclusions of highly limited applicability. It is the purpose of this paper to discuss, in a general fashion, the importance of the "mechanical" effects and to describe how they are related to rheological properties. Since the rheology of non-Newtonian fluids and the techniques required to study it are not discussed widely in the biotechnology literature, we begin with a concise review of this topic.

2. Rheology: Principles and Methods

2.1. Basic Concepts and Definitions; Newtonian Fluids and Laminar Flow

The movement of solid surfaces (e.g., an impeller) in contact with a fluid causes the fluid to move in some characteristic pattern which results in the development of internal stresses and the application, on the solid surfaces, of characteristic forces which must be continuously counterbalanced (e.g., by a drive motor) in order to sustain the fluid motion. The nature of the flow pattern and the magnitudes of internal stresses and applied forces depend primarily on the geometry of the system, the rate of fluid motion, and the intrinsic rheological (flow) properties of the fluid. A very simple example illustrating these phenomena is given in Fig. 1. Here a liquid is confined between two coaxial cylinders of large enough radii that curvature is negligible and exerts no influence; the surfaces approximate parallel flat planes. From experience we know that if we rotate the inner cylinder the fluid will move and a torque will have to be applied to the outer cylinder to keep it stationary. Clearly, then, the fluid exerts a tangential force on the outer cylinder while the inner cylinder exerts such a force on the fluid in contact with it and this force is transmitted through the fluid from one layer to the next. At any point in the fluid, the tangential force divided by the surface area on which it acts is defined as the *shear stress*. Experience teaches that for a given fluid the magnitude of this shear stress varies with the rate of fluid motion (which is determined in part by the speed of the driven cylinder). Furthermore, if the rate of rotation is not too high the only macroscopic component of fluid velocity will be in the direction of bulk flow; as shown in Fig. 1 the velocity component in both the r or θ directions will

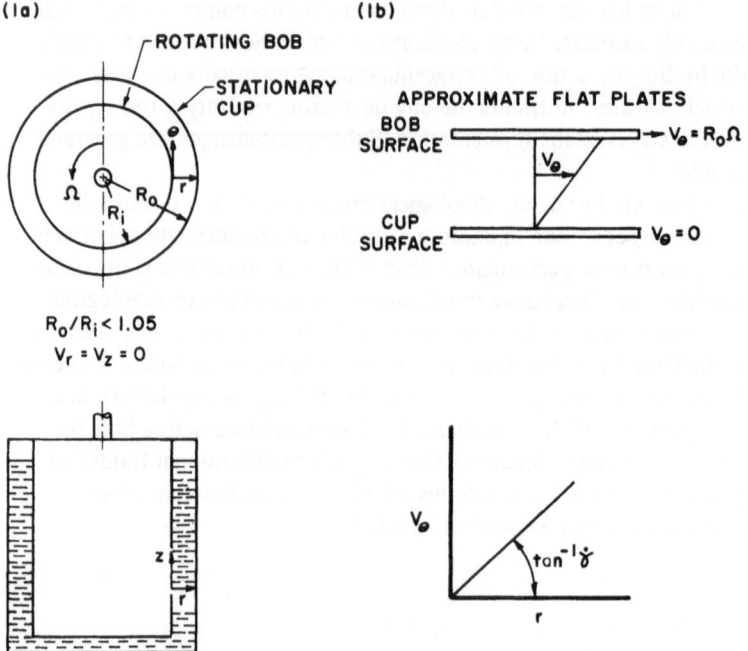

Fig. 1. (a) Laminar flow between concentric cylinders having negligible curvature. (b) Velocity profile; flatt plate approximation

be zero. This type of fluid motion is called *laminar flow:* A dye stream injected at any point will move in a straight line and the only dispersion will be that due to random (thermal) molecular motion.

Analysis of the fundamental equations [2] of fluid dynamics reveals that for laminar flow *of the type illustrated in Fig. 1,* the shear stress will be constant throughout the fluid and the fluid velocity will very linearly from zero at the fixed surface to $R_0\Omega$ at the rotating surface. To further describe the nature of the flow field we define, for laminar flow, the shear rate, $\dot\gamma$, as

$$\dot\gamma = \frac{du}{dn} \tag{1}$$

where n represents the direction perpendicular to the direction of the fluid motion at a point and u is the fluid velocity. Shear rate is usually reported in units of s^{-1}. (More generalized definitions are discussed in various works on fluid mechanics and rheology [2, 3].) For the case under consideration

$$\dot\gamma = \frac{du}{dx} = \frac{\Delta y}{\Delta x} = \frac{u_i}{\Delta x} \tag{2}$$

where u_i is the velocity of the inner cylinder and Δx is the thickness of the liquid layer. It must be emphasized that Eq. (2) is valid only for the flow illustrated in Fig. 1 (for example, it is not valid for coaxial cylinders for which curvature does effect the flow pattern).

For a variety of fluids (e.g., water, sugar solutions) it is found that the relationship between shear stress, τ, and shear rate (the constitutive equation) for laminar flow is given by

$$\tau = -\eta\dot{\gamma} \tag{3}$$

where η is a constant called the fluid viscosity and is generally reported in units of centi-poises (1 centipoise $= \dfrac{1\ g}{cm \cdot s}$). [The minus sign in Eq. (3) is a convention.] Such fluids are called *Newtonian* fluids.

Except for special cases which need not concern us here, Eq. (3) is valid for all simple laminar flows of Newtonian fluids. However, it must be recognized that the expression for shear rate, $\dot{\gamma}$, is not always as simple as that for flow between parallel flat plates because the velocity profile (flow pattern) will not always vary in a simple linear fashion. For example, it can be shown [2] that when a Newtonian fluid flows in laminar fashion through a round tube, the velocity varies parabolically from zero at the wall to a maximum at the center of the tube;

$$u_z = u_m \left[1 - \frac{r}{R}\right]^2 \tag{4}$$

where

$$
\begin{aligned}
u_z &= \text{fluid velocity at radial position } r \\
u_m &= \text{maximum fluid velocity at } r = 0 \\
R &= \text{tube radius}
\end{aligned}
$$

We can use Eq. (4) to calculate the fluid shear rate;

$$\dot{\gamma} = \frac{du}{dn} = \frac{du_z}{dr} = -\frac{2u_m}{R}\left(\frac{r}{R}\right) \tag{5}$$

The shear stress is then

$$\tau = -\eta\dot{\gamma} = \frac{2u_m}{R}\ \eta\left(\frac{r}{R}\right). \tag{6}$$

Thus, both the shear rate and shear stress very linearly from zero at the center of the tube to maximum values at the tube wall. As will be discussed in Sect. 2.4.1., both shear rate and shear stress at the tube wall are easily calculated (for Newtonian fluids) from measurements of flow rate and pressure drop and therefore tube or capillary flow provides a convenient means for determining Newtonian viscosity in many cases.

For coaxial cylinders which do not approximate parallel flat plates it can be shown that
for Newtonian fluids in laminar flow, the shear rate is [4]

$$\dot{\gamma} = -r\frac{d\omega}{dr} = -\frac{2\Omega}{r^2}\frac{R_i^2 R_o^2}{R_o^2 - R_i^2}$$ (7)

and the shear stress is [5]

$$\tau = \frac{M}{2\pi r^2 h}$$ (8)

where

ω	=	angular velocity
Ω	=	angular velocity of inner cylinder
R_i	=	radius of inner cylinder
R_o	=	radius of outer cylinder
M	=	torque
h	=	liquid height.

(9)

The coaxial system can therefore be used in a simple fashion to determine the viscosity
of Newtonian fluids since

$$\eta = \frac{\tau}{\gamma} = \frac{M_0}{4\pi h \Omega}\left(\frac{1}{R_o^2} - \frac{1}{R_i^2}\right).$$ (10)

(For further discussion of the coaxial cylinder viscometer, including various correction
factors, see Skelland [6], Middleman [7] and the series of articles by Krieger and Maron
[8, 9, 10]).

2.2. Non-Newtonian Fluids

2.2.1. Pseudoplasticity

Many fluids, including various culture broths, do not exhibit Newtonian behavior.
Among the more common non-Newtonian characteristics is *pseudoplasticity*. A typical
flow curve (shear stress *vs* shear rate) of a pseudoplastic fluid is illustrated in Fig. 2a.
The ratio of shear stress to shear rate in a well-defined laminar flow field is called the
apparent viscosity, η_a, and decreases with increasing shear rate as shown in Fig. 2b:

$$\eta_a = \frac{\tau}{\gamma}.$$ (11)

The apparent viscosity is, in a sense, an equivalent Newtonian viscosity but generally is
not equal to the slope of the flow curve. Usually, pseudoplastic fluids exhibit "New-

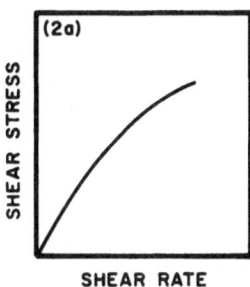 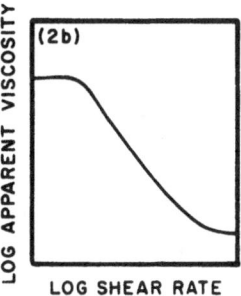

Fig. 2. Pseudoplastic fluid (a) flow curve and (b) viscosity characteristics

tonian" behavior at very low shear rates (zero shear viscosity) and at very high shear rates (infinite shear viscosity). In any event, it should be clear it is meaningless to specify the viscosity of a pseudoplastic fluid without noting the shear rate at which it was measured. It must be emphasized that the above definition of viscosity (which is accepted by rheologists) requires measurement with an apparatus that provides laminar flow and permits unambiguous calculation of both shear rate and shear stress. It should be noted also that expressions for shear rate and shear stress are much more complex for pseudoplastic fluids than for Newtonian fluids even for simple geometries. This complicates the measurement of viscosity and will be discussed at greater length in Sect. 2.4.2.1.

Pseudoplastic behavior results in flow characteristics which depart significantly from those of Newtonian fluids. Characteristics of laminar flows through round tubes are illustrated in Fig. 3. In comparison to Newtonian flow, the pseudoplastic flow is typified by a relatively flat velocity profile with significant gradients only near the wall and hence shear rates which are rather high near the wall and very low elsewhere. As a result viscosity is relatively low near the wall and relatively high elsewhere. Clearly, this has a significant effect on both mass and heat transfer. Similar behavior is observed in stirred

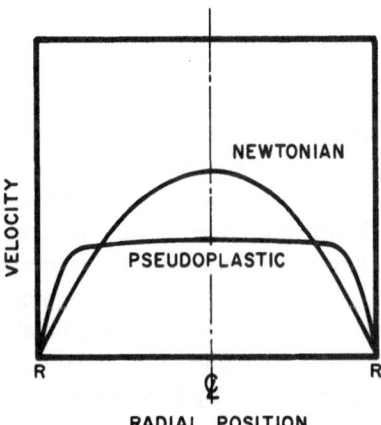

Fig. 3. Velocity profiles for Newtonian and pseudoplastic fluids in laminar flow through a round tube

tanks; shear rates are usually high very near the impeller and low elsewhere and hence fluid viscosity is low near the impeller but high at even relatively short distances from it [11]. This behavior of pseudoplastic fluids is illustrated in Fig. 4 [11]. Again, the effects on mass and heat transfer are quite important.

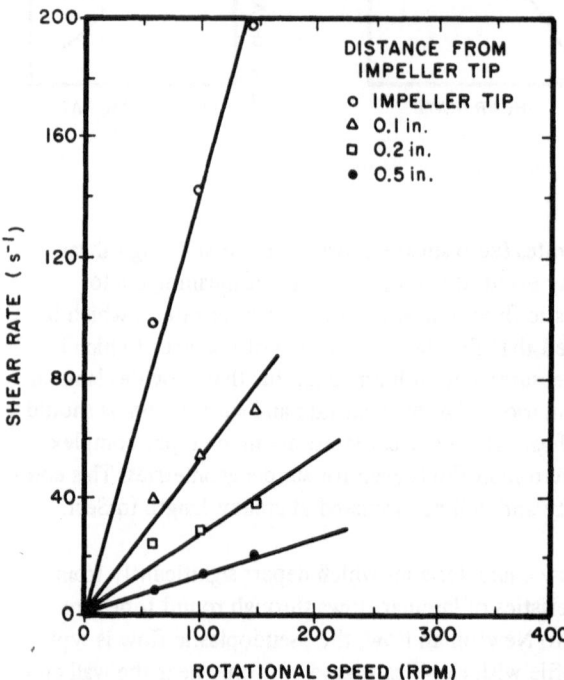

Fig. 4. Shear rate *vs.* rotational speed at various distances from the tip of a turbine impeller [11]

A variety of mathematical expressions has been developed to correlate viscosity data for pseudoplastic fluids. None is completely satisfactory for all fluids nor for the entire range of shear rates for any given fluid. The *Power Law* model is probably the most widely used because of its simplicity:

$$\eta = K\dot{\gamma}^{m-1} \qquad (12)$$

Where K is the *consistency index* (given in pseudopoise) and m is the *power law index*. In general m and K are constant only over limited ranges of shear rate. For pseudoplastic fluids m is less than 1; the smaller m the more rapid is the decline of viscosity with increasing shear rate and the more pronounced are the effects of pseudoplasticity on flow and other transport phenomena. K is a direct measure of viscosity at a given rate of shear; the larger K, the greater the viscosity at a given rate of shear. However, it must be recognized that, in general, particular values of K and m are valid only for some particular range of shear rates. Furthermore, the value of K is obtained by extrapolation of viscosity data to 1 s^{-1} and may not have real physical significance at that rate of shear.

Some of the other mathematical models for pseudoplastic fluids which have been used widely are [12]

(a) Eyring

$$\tau = \frac{\dot{\gamma}}{C_1} + C_2 \sin{(\tau/C_3)}. \tag{13}$$

(b) Powell-Eyring

$$\tau = C_1 \dot{\gamma} + C_2 \sinh^{-1}{(C_3\gamma)}. \tag{14}$$

(c) Prandtl

$$\tau = C_1 \sin^{-1}{(\dot{\gamma}/C_2)}. \tag{15}$$

(d) Williamson

$$\tau = \frac{C_1 \dot{\gamma}}{C_2 + \dot{\gamma}} + C_3 \dot{\gamma}. \tag{16}$$

These are usually more difficult to apply than the power law model and have not been found useful for culture broths. More extensive discussions of these and other models along with more generalized discussions of mathematical descriptions of shear rate can be found in a variety of works on fluid mechanics and rheology [13–16].
Finally, fluids whose viscosities increase with increasing shear rate are called *dilatant*. Such behavior has not been reported for culture broths.

2.2.2. Yield Stress

Some fluids will not flow until some minimum shear stress (*yield stress*) is exceeded. Fluids which exhibit Newtonian-like behavior once flow is initiated (i.e., once the yield stress is exceeded) are called *Bingham plastics* (see Fig. 5). Bingham plastic behavior can be expressed mathematically as [17]

$$\tau - \tau_y = \eta_\rho \dot{\gamma} \qquad \tau > \tau_y, \tag{17}$$

where τ_y is the *yield stress* and η_ρ is the plastic *viscosity* or *rigidity*.

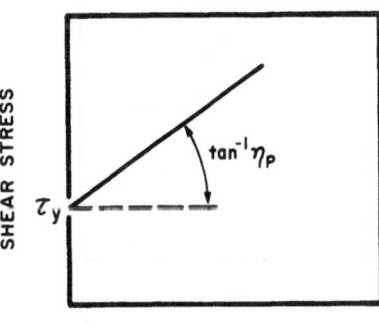

Fig. 5. Bingham plastic flow curve

A number of culture fluids exhibit yield stress but not all behave as Bingham plastics. Most of the reported cases involve fungal cultures in which yield stresses are associated with the structure imposed by the mycelial mat and the breakdown of this structure upon application of a stress greater than the yield stress. Experimental data for such systems will be presented in Sect. 3.1. but it is worth noting here that Roels *et al.* [18] have developed a rheological model which incorporates explicitly mycelial structure and which is based on the Casson viscosity equation:

$$(\tau)^{1/2} = (\tau_y)^{1/2} + K_c(\dot{\gamma})^{1/2}. \tag{18}$$

These authors found that the Casson equation, which incorporates features of the power law and the Bingham plastic equations, is superior to either in correlating experimental viscosity data for two different strains of *P. chrysogenum.*

The effects of yield stress on flow are somewhat similar to those of pseudoplasticity but are much more pronounced. For example, when a Bingham plastic flows through a round tube (laminar flow) it is quite possible for a "solid" plug of fluid to exist from the center of the tube to a point where the shear stress is just exceeded. This is illustrated in Fig. 6 [17]. Similar effects can occur in agitated vessels.

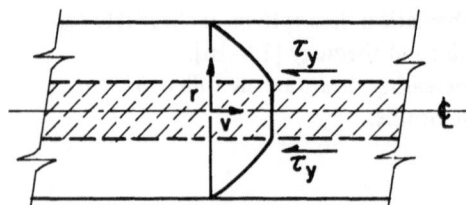

Fig. 6. Velocity profile for a Bingham plastic in laminar flow through a round tube [17]

2.2.3. Time Dependent Viscosity

Fluids whose viscosities increase or decrease with time of shearing are called *rheopectic* and *thixotropic* respectively [19]. Thixotropic behavior is illustrated in Fig. 7a [19]. When thixotropic (or rheopectic) fluids are subjected to varying shear rate they exhibit

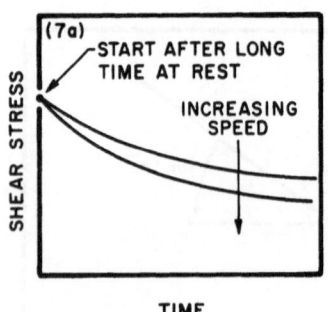
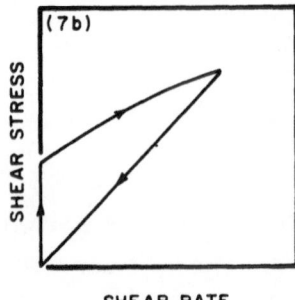

Fig. 7. Thixotropic fluid (a) time-dependent shear thinning, (b) hysteresis [19]

hysteresis as illustrated in Fig. 7b. Thixotropic behavior appears to be associated with reversible structure effects and could be important in mixing cultures containing fungal mycelia or extracellular microbial polysaccharides. As already noted, (Sect. 2.2.1.) a culture fluid is exposed to various shear rates as it moves about the reactor and might therefore exhibit viscosity hysteresis. The importance of this effect will depend on the time scale of the structure effects compared to the time scale of mixing and on the importance of thixotrophy in comparison to other phenomena such as pseudoplasticity. To the author's knowledge there have been no reports of any significant efforts to determine the importance of thixotropy (or rheopexy) in culture broths.

Finally, it should be noted that even if thixotropic properties are not important during the reaction *per se*, they can influence the outcome of an experiment designed to study power consumption [20, 21]. When performing such experiments the investigator should always keep in mind that (1) at a given impeller speed power uptake may vary with time and (2) the power consumption at a given speed may depend on whether the preceeding speed was higher or lower and on the time scale of the experiment.

2.2.4. Viscoelasticity

Many fluids, particularly solutions of high polymers such as microbial polysaccharides [20], exhibit elastic response superimposed on a characteristic viscous behavior and are therefore called *viscoelastic*. If the fluid contained in the simple rheometer illustrated in Fig. 1 is non-elastic and we uncouple the driven cylinder from its drive shaft we will find that motion continues in the original sense for the short time required to dissipate the kinetic energy of the system. However, if the fluid is viscoelastic, the driven cylinder will reverse its direction. This is caused by elastic stresses developed during the steady state flow period. These rather complex stresses are responsible for a variety of phenomena including the well-known Weissenberg effect illustrated in Fig. 8. Most viscoelastic fluids are pseudoplastic and may exhibit other rheological characteristics such

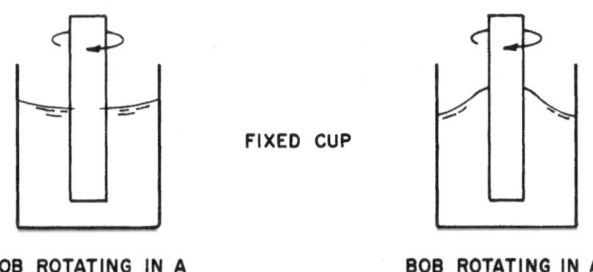

Fig. 8. Rod-climbing (Weissenberg) effect

FIXED CUP

BOB ROTATING IN A PURELY VISCOUS FLUID

BOB ROTATING IN A VISCOELASTIC FLUID

as yield stress. Therefore, mathematical expressions for viscoelastic behavior are generally very complex and are not suitable for use in analysis of complex systems such as bioreactors. (A comprehensive treatment of viscoelastic behavior has been given by Frederickson [15].) Finally, viscoelastisticity has been shown to affect flow patterns in agitated vessels [22] but the importance of this effect during cultivation has not been studied.

Before concluding this section it should be re-emphasized that non-Newtonian behavior results in flow patterns (some of which were described above) which depart markedly from those exhibited by Newtonian fluids and that the nature and magnitude of such departures are influenced profoundly by system geometry and operating conditions. In turn, these effects on flow pattern affect all of the transport phenomena and therefore must be considered not only for the reaction *per se* but also for subsequent recovery operations.

2.3. Turbulent and Other Complex Flows

Thus far we have made a point of confining our attention to laminar flow. This has been done to insure an unambiguous definition of viscosity. When the flow becomes rapid enough or the fluid is subjected to accelerations (e.g., rapid change of direction) severe enough to establish components of velocity perpendicular to the main flow then a turbulent flow regime is established and a precise definition of shear rate becomes quite difficult. In addition, the magnitude of the shear stress is no longer simply related to shear rate and viscosity (a result of momentum transfer via turbulent eddies). Viscosity calculated on the basis of turbulent flow measurement varies with flow conditions even for Newtonian fluids and hence such measured values can not be thought of as intrinsic fluid properties. Indeed, the development of reliable theoretical expressions for such simple calculations as pressure drop *vs.* flow rate for turbulent flow of a Newtonian fluid through a round pipe is quite difficult and as a result the accepted practice in designing for turbulent flow has been to employ empirical methods based on dimension-less parameters of proven value in providing reasonably reliable scale-up correlations. (The same approach is often used for laminar flow in complex geometries such as agitated vessels.)

One of the most important of these dimensionless parameters is the *Reynolds number*, N_{Re}, which, in essence, represents the ratio of inertial forces (attributed to fluid momentum) to viscous forces. It is useful not only for purposes of correlation and scale-up but also for delineating laminar and turbulent flow regimes. The most useful form of the N_{Re} depends on the system geometry. Thus, for round tubes,

$$N_{Re} = \frac{D\bar{u}\rho}{\eta}, \tag{19}$$

where D is the tube diameter, $\bar{u}$ is the average fluid velocity, ρ the fluid density and η the fluid viscosity, while for agitated vessels

$$N_{Re} = \frac{d_i^2 N\rho}{\eta} \tag{20}$$

where N is the rotational speed of the impeller and d_i is the impeller diameter. Application of these definitions is simple and straightforward for Newtonian fluids but somewhat involved for non-Newtonian fluids: In particular, it is not clear immediately as to what value of viscosity should be used. However, a number of investigators [23, 27] have developed and tested empirical methods which provide an adequate resolution of this difficulty for many situations of interest to the biotechnologist. In particular, Metzner and his coworkers [11, 23] have shown that for many non-Newtonian fluids, the shear rate at the impeller tip in an agitated vessel is given by

$$\dot{\gamma} = kN, \tag{21}$$

where k may be taken as approximately 10 for turbine impellers, at least for many pseudoplastic non-Newtonian fluids. A more extensive compilation of k values for various impeller types and for dilatant and Bingham fluids can be found in the extensive reports of Calderbank and Moo-Young [25, 26]. In any event, useful generalized correlations [24–27] for power input have been developed by calculating N_{Re} (consult original references for precise forms of N_{Re} to be used) on the basis of a viscosity evaluated at the shear rate specified by Eq. (21) and then plotting it against a dimensionless power number, N_P, given by

$$N_P = \frac{P}{\rho N^3 d_i^5} \tag{22}$$

where P is the power input.

It is important to recognize that the flow in an agitated vessel is not simple laminar flow and that the shear rate given by Eq. (21) is a simplified approximation. Furthermore, the impeller-tip shear rate is usually much greater than shear rates elsewhere in the vessel and therefore can give one a false impression of the overall mixing intensity. The impeller tip shear rate certainly has proven valuable for correlation purposes but it does have shortcomings and work to find a better measure of flow behavior would be justified.

2.4. Measurement of Rheological Properties

Many useful instruments are available for the measurement of rheological properties. However, it must be recognized from the outset that not all of the popular instruments (e.g., Brabender Amylograph) measure viscosity as we have defined it, even though they provide valuable flow behavior information which, in many cases is quite adequate so long as it is internally consistent. (In some cases such information is even more useful than actual viscosity data.) In general, the complex flow patterns employed in many instruments makes them unsuitable for rigorous measurement of viscosity as we have defined it. It is also important to note that even with instruments specifically designed to measure viscosity, considerable care must be exercised in analyzing data for non-Newtonian fluids (see Sect. 2.4.1.). It is also important to recognize that in most cases the empirical constants in any viscosity model are valid over only a limited range of shear rate and extrapolation beyond the range for which the constants were determined usually is risky. For example. the consistency index, K, of the power law model is obtained by extrapolation to $1 \ s^{-1}$. Therefore, it is quite possible that K will have no real physical significance at $1\ \mathrm{s}^{-1}$. Also, the value of K will generally vary with the range of shear rates considered. Similar statements can be made regarding the flow index, m. Thus, viscosities should be determined over the range of shear rates that will be encountered in the process of interest. Finally, it should be noted that the choice of the most appropriate viscometer will be dictated, in part, by the range of shear rate to be studied.

As space will permit only a brief discussion of this most important topic of rheological measurements, the reader is referred to the reviews and treatises on rheological techniques [28, 29].

2.4.1. Capillary Viscometers

In its simplest form the capillary viscometer comprises a true-bore capillary (of accurately known diameter and length) attached to a pressurized reservoir containing the test fluid. The entire system is accurately thermostated. To determine the viscosity of a simple Newtonian fluid one must measure only the pressure drop across the capillary and the volumetric flow rate since the shear rate and shear stress for laminar flow are easily calculated from [2]

$$\dot{\gamma}_w = \frac{4Q}{\pi R^3} \tag{23}$$

and

$$\tau_w = \frac{\Delta P \cdot R}{2L} \tag{24}$$

where $\dot{\gamma}_w$ and τ_w are the shear rate and shear stress respectively at the capillary wall, Q is the volumetric flow rate, ΔP is the pressure drop, and L and R the length and radius respectively of the capillary. Thus

$$\eta = \frac{\tau_w}{\gamma_w} = \frac{\Delta P \cdot \pi R^4}{8LQ} \tag{25}$$

Eq. (25) is frequently called the Hagen-Poiseuille equation. For accurate work a number of instrument corrections should be evaluated [30–32].

For non-Newtonian fluids the situation is complicated by the fact that the shape of the velocity profile and hence the shear rate are dependent on the specific non-Newtonian characteristics of the fluid being tested. To calculate non-Newtonian viscosity from capillary data one can use the Rabinowitsch-Mooney equation [33, 34] (valid for any fluid),

$$\dot{\gamma}_w = \frac{4\,Q}{\pi R^3} \left\{ \frac{3}{4} + \frac{1}{4} \frac{d \log Q/\pi R^3}{d \log \Delta P \cdot R/2L} \right\} \tag{26}$$

to calculate the shear rate at the wall and Eq. (24) to calculate the shear stress at the wall. However, one must measure the volumetric flow rates for various applied pressures in order to evaluate the derivative in Eq. (26). Special techniques for data reduction are available for fluids which conform to Bingham plastic or power law models [35, 36]. For a discussion of various corrections (e.g., for heat effects, viscoelasticity, and effects) consult Van Wazer et al. [35].

It is frequently difficult to use a capillary viscometer to measure viscosities of fluids containing suspended particles. Among the problems most frequently encountered are settling of particles and plugging of fine capillary tubes, particularly by filamentous fungi. When particles remain suspended and do not clog the capillary, complications can be introduced by the "tube pinch" effect [37] which is the migration of small particles away from the wall to the center of the capillary. Capillary rheometers are also not well suited for measurement of yield stress. As a result of these difficulties capillary viscometers have not been used widely to study rheological properties of culture broths. However, their usefulness in characterizing clarified fluids should not be overlooked.

2.4.2. Rotational Viscometers

2.4.2.1. Coaxial Cylinder (Couette) Viscometer

The coaxial cylinder instrument discussed in Sect. 2.1. is an example of a classical rotational viscometer which provides a straightforward and accurate means for determining Newtonian viscosity as already discussed. While it can also be used for determining non-Newtonian viscosity, somewhat involved calculations are required because the shear rate is not related simply to rotational speed and geometric factors as it is for Newtonian fluids [Eq. (7)]. The calculations are similar to those required for solution of the Rabinowitsch equation [Eq. (26)] and require the collection of data at several rotational speeds. For details regarding calculations, experimental procedures and various instrument correction factors consult Van Wazer et. al. [28] and Krieger and Maron [8–10]. Finally, it should be noted that when suspensions are tested in the Couette viscometer or any other rotational viscometer having a smooth rotating bob, a clear region forms in the immediate vicinity of the bob [38] and there is often noticeable gravity settling of the particles. Hence, some question exists as to whether such viscometers are entirely appropriate for testing whole culture broths (see Sect. 2.4.2.4.). Further investigation is required.

2.4.2.2. The Brookfield Viscometer

A widely used variation of the coaxial viscometer is the Brookfield viscometer, illustrated in Fig. 9 [39]. With the guard legs removed, this rugged and versatile instrument can be used to closely simulate a rotating cylinder in an "infinite" medium. For this case the shear rate at the rotating cylinder surface is given with reasonable accuracy by

$$\dot{\gamma} = \frac{4\pi N}{n^1},$$
(27)

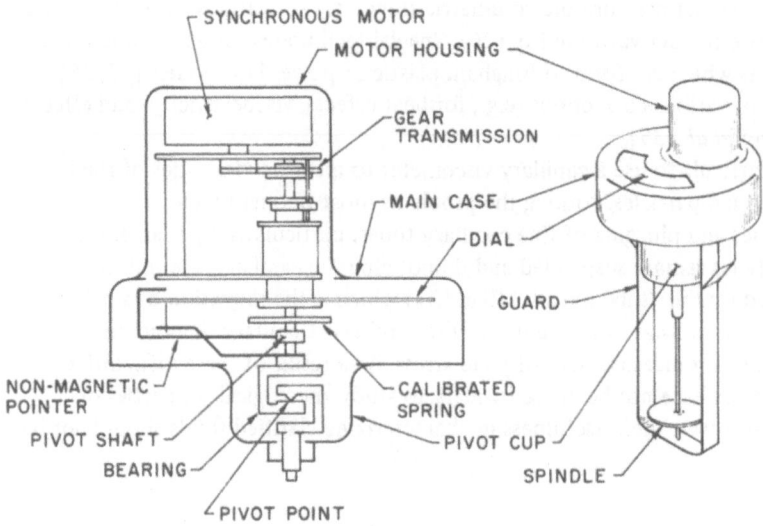

Fig. 9. Brookfield viscometer

where n^1 is the slope of the logarithmic plot of torque *vs* rotational speed [40]. The only caution that need be exercised is to insure that the container holding the sample is large enough to eliminate stationary-surface effects. The shear stress is still given by Eq. (8). Finally, it should be noted that this instrument (1) exhibits phase separation near the bob and gravity settling of particles when suspensions are tested and (2) is not recommended for determination of yield stress of solutions or suspensions.

2.4.2.3. Cone and Plate Viscometer

Another widely used type of rotational instrument is the cone and plate rheometer. This device, which is illustrated in Fig. 10, has the very convenient feature that for cone

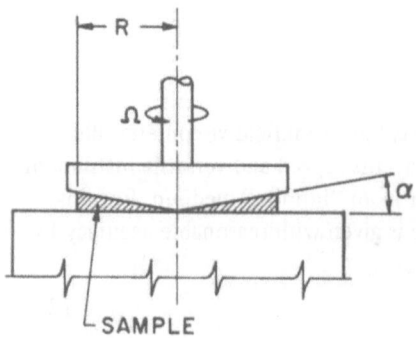

Fig. 10. Cone and plate viscometer

angles less than 3°, the shear rate throughout the fluid is essentially constant and given by [41]

$$\dot{\gamma} = \frac{\Omega}{\alpha},$$
(28)

where Ω is the rotational speed of the bob and α is the cone angle (in radians). The shear stress (at the surface of the bob) is given by [41]

$$\tau = \frac{3M}{2\pi R^3},$$
(29)

where M is the torque exerted on the bob. Thus, non-Newtonian flow curves can be obtained directly with no tedious calculations. While the cone and plate has been used extensively to study non-Newtonian fluids in general, its use in the study of rheological properties of culture broths has been exceedingly limited. Finally, the author has observed phase separation near the rotating member when suspensions were tested, but the separation did not appear to be as pronounced as in the cases of Couette and Brookfield viscometers.

2.4.2.4. "Turbine" Viscometer

While standard rotational viscometers are versatile, reliable, and widely employed, they are sometimes difficult to use for studying fermentation broths, particularly mycelial broths. Some of the difficulties that have been enumerated by Bongenaar et al. [42] and by Roels et al. [18] are:
1) phase separation; formation of less dense region in immediate vicinity of rotating bob
2) non-homogeneity as a result of gravity settling
3) large particles about the same size as the gap impair accurate measurement of viscosity
4) destruction of particles in shear field.

A typical example of the difficulties encountered is illustrated in Fig. 11 [43]. These data, taken in the author's laboratory, represent Brookfield viscometer (model LVT) scale readings (proportional to viscosity) as a function of time at various speeds for a 1% (w/v) Aspergillus niger culture broth sample. The readings are erratic but tend to decrease with time (more so at higher speeds) and at first seem to indicate thixotropic behavior. However, visual observation reveals partial settling and also phase separation near the rotating spindle (No. 1) are responsible for the observed time effect. It is interesting to note that similar behavior is exhibited by yeast suspensions [~ 4% (w/v)], although to a much smaller extent [44].

In order to overcome such difficulties Bongenaar et al. [42] and Roels et al. [18] have developed a modified viscometer (basic unit was a Haake Rotovisko; MK50) which employs a turbine impeller instead of the more conventional smooth, geometrically simple bobs. They claim that this provides two important advantages:

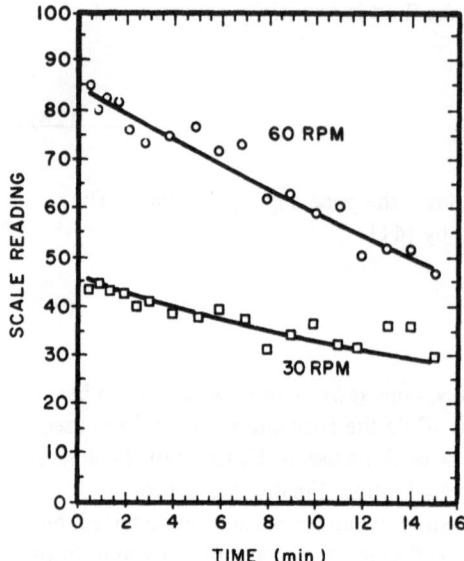

Fig. 11. Effect of phase separation on Brook-field viscometer reading; 1% *A. niger* culture broth

a) phase separation and settling are prevented, and

b) the shear rate is simply related to the impeller speed via Eq. (21).

The turbine impeller establishes a relatively complex flow pattern which does not allow straightforward calculation of shear rate and raises some questions regarding the significance of the calculated viscosity. In particular, the instrument does not seem to satisfy the requirements for measuring viscosity as defined formally and therefore it is not clear immediately that the calculated quantity is an intrinsic property of the fluid. In this regard it is not unlike "practical" viscometers, such as the Stormer [45] viscometer, which are used to provide practical and important flow properties which do not necessarily correlate directly or simply with viscosity. (Indeed there are cases in which such data is more valuable than viscosity *per se*.) Nevertheless, the instrument does eliminate operating problems associated with rotational viscometers used to study mycelial broths and appears to give internally consistent data. Furthermore, data analysis while not absolutely rigorous from a rheological point of view, is based on well-proven and widely accepted empirical correlations. Roels *et al.* [18] have presented the following derivation. The generally accepted empirical relationship [24] between power number, N_P and Reynolds number, N_{Re}, in the laminar flow region ($Re < 10$) is

$$N_P = \frac{64}{N_{Re}} \tag{30}$$

The power number is, by definition,

$$N_P = \frac{P}{\rho N^3 d_i^5} \tag{31}$$

where P is the power input. The power input is also related to the torque, M, exerted on the impeller;

$$P = 2\pi NM. \tag{32}$$

Combining Eqs. (30) through (32) and the defining equation for the Reynolds number one obtains

$$M = \frac{64\,d_i^3}{2\pi}\,\eta_a. \tag{33}$$

As stated previously, it has been shown experimentally [23] that the shear rate at the tip of a turbine can be expressed conveniently and usefully as

$$\dot{\gamma} = kN \tag{34}$$

where k is a constant. (The N_P–N_{Re} correlations for non-Newtonian fluids were developed by calculating N_{Re} on the basis of viscosity evaluated at the impeller tip shear rate. See original papers [26–28] for details.) Combining Eq. (34) with the defining equation for viscosity results in

$$\tau = \dot{\gamma}\,\eta_a = \eta_a\,kN. \tag{35}$$

Finally, Eqs. (33) and (35) are combined to give

$$\tau = \frac{2\pi k}{64\,d_i^3}M \tag{36}$$

or

$$\tau = C_i\,M \tag{37}$$

where C_i may be viewed as an instrument constant. Thus one may determine viscosity as a function of shear rate by simply measuring torque at various rotational speeds, so long as $Re < 10$.

While the method does offer considerable promise for the study of mycelial broth rheology, it will definitely require further testing. In particular, comparisons should be made between viscosities of homogeneous non-Newtonian fluids determined by the new method and by conventional methods. Also, effects of gas bubbles and of various geometric parameters (e.g., impeller size and configuration, presence of baffles) should be investigated carefully.

2.4.2.5. Viscosity Measurement in situ

It has been observed by a number of biotechnologists [46, 49] that it would be most advantageous (particularly from the point of view of process control) to measure rheological properties in situ during a bioprocess. Wang and Fewkes [48] have attempted to accomplish this goal by using a technique similar to Bongenaar's method to measure the rheological properties of a *Streptomyces niveus* culture *in situ*. They state that this can be accomplished by reducing the agitator speed to ensure "laminar" flow and then measuring power input as a function of agitator speed. (The necessity for maintaining laminar flow when applying Bongenaar's method is clear since the calculations are based on the assumption of laminar flow. However, the necessity for laminar flow in the present case is not clear.) While they give no details of the calculation it is probably safe to assume that their measurements were used in conjunction with the N_P vs. N_{Re} correlation of either Metzner and Otto [24] or Calderbank and Moo-Young [25, 26] and Eq. (34) for shear rate at the impeller tip to obtain the relationship between viscosity and shear rate. (The N_P–N_{Re} correlations for non-Newtonian fluids were developed by calculating N_{Re} on the basis of viscosity evaluated at the impeller tip shear rate. See original papers [24–26] for details.) As an example, one can proceed as follows:

(1) Calculate $N_P \left(= \dfrac{P}{\rho N^3 d_i^5} \right)$.

(2) From the Methner-Otto [31] correlation obtain the corresponding N_{Re}.

(3) Calculate the apparent viscosity, $\eta_A = \left(\dfrac{\rho N d_i^2}{N_{Re}} \right)$.

(4) Calculate the shear rate at the impeller tip, $\hat{\gamma} \, (= kN)$.

For other, but essentially similar, approaches the reader is referred to Skelland's text [26].

Wang and Fewkes [48] state that for the *S. niveus* process studied, power law behavior is exhibited at any given cell mass concentration. However, as can be seen from some of their results, which are given in Fig. 12, [48] both m and K appear to be functions of

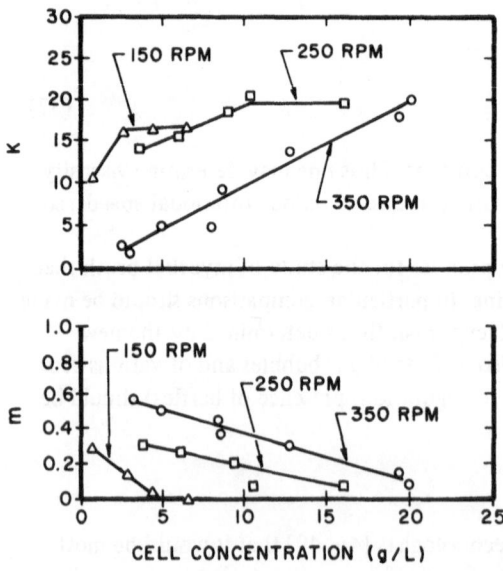

Fig. 12. Viscosity characteristics of *S. niveus* culture broth measured *in situ* [48]

rotational speed and this somewhat disturbing. In particular, since the agitation speed varies from 150 RPM to 350 RPM the shear rate at the impeller tip varies between approximately 25 s^{-1} and 75 s^{-1}, only one third of a decade. For such a narrow range, one would expect the power law model to correlate the data reasonably well with only one value each for m and K. The fact that m and K do vary leads one to believe that there may have been some deficiency in the method *as used*. The system geometry might have been such as to cause minor but significant deviations from the standard N_P–N_{Re} correlations. These correlations do exhibit some variability and the sensitivity of viscosity calculations to such variations was not determined. Failure to shut off air flow during rheological measurements (it is not clear from the paper if this was the case) would also cause discrepancies since the power number–Reynold's number correlation is distorted markedly by normal air flow rates. Entrapped air bubbles might also have caused problems. Despite these problems, the method is quite promising and should be investigated further in depth. As for *in situ* instruments, it should be noted that at present there appears to be only one commercial viscometer suitable for use in agitated vessels. This is the Dynatrol (Automation Products, Inc.) which has been used for various process applications. However, no reports have been published of its successful use in fermenters.

3. Rheological Data for Culture Fluids

3.1. Mycelial Culture Broths

The rheological properties of mycelial broths are usually controlled by the concentration and morphological state of the mycelia. The only important exception to this occurs when starting media contain either high levels of undissolved solids or high concentrations of high molecular weight substrates such as corn starch. However, even in these cases, the importance of medium constituents generally decreases as cell mass increases (e.g., hydrolysis of starch).

Morphological conditions should exert a very profound effect on the nature of broth rheology. One would expect that disperse filamentous growth should tend to "structure" the entire suspension (see Roels *et al.* [18], for a more fundamental discussion) and cause yield stress, pseudoplasticity, or both. On the other hand, pellet-like growth should lead to more Newtonian-like behavior and lower Viscosity [49] although some degree of non-Newtonian response should be expected because the pellets are quite deformable.

These arguments seem to be supported by the majority of the published data summarized in Table 1 and Fig. 13–17. However, there is disagreement, difficulty in comparing results of different investigators, and an unacceptably high level of ambiguity which makes it difficult to reach firm conclusions or to establish useful generalities. Among the more important reasons for these difficulties are:

Table 1. Rheological properties of mycelial culture fluids

Culture	Shear rate[a] (s^{-1})	Viscometer	Flow curves given	Comments
A. niger [50] (washed cells)	Not given	Ferranti VM (concentric cylinder)	No	1. Bingham plastic behavior mentioned, but not quantified. 2. Gives K values for various cell mass concentration; $K = 2.65 \log C(gm/l)$
P. chrysogenum [51] (whole broth)	1–10	MacMichael (concentric cylinder)	Yes	1. Flow curves demonstrate Bingham plastic behavior. 2. τ_y & η_p vary directly with cell mass concentration and/or fermentation time. See Figs. 13 and 14.
P. chrysogenum [52] (whole broth)	< 0.2[b]	Brookfield (guard removed)	Yes	1. Flow curve (Fig. 15) seems to indicate both yield stress and some other pronounced structure effect. 2. Instrument artifact or error in calculation of γ suspected.
P. chrysogenum [42] (whole broth)	0.1–1.5	"Turbine"	Yes	1. Casson behavior observed. 2. Yield stress and Casson viscosity, η_c, increase with reaction time and cell concentration.
P. chrysogenum [18] (whole broth and diluted whole broth)	< 15	"Turbine"	Yes	1. Observations parallel those of preceeding example. 2. Model incorporating morphological concepts developed.
Endomycopses sp. [55] (whole broth)	not given	Ferranti coaxial cylinder	No	1. Power law behavior. 2. K and m varied appreciably but not monotonically over course of batch (see Fig. 15).
C. hellebori [52] (whole broth)	< 0.2[b]	Brookfield (guard removed)	Yes	1. Bingham plastic behavior early and late in batch. 2. Yield stress (equal to Bingham yield stress late in batch and pseudoplasticity at intermediate times.
S. griseus [52] (whole broth)	< 0.2[b]	Brookfield (guard removed)	Yes	1. Bingham plastic behavior early in fermentation. 2. Newtonian behavior late in batch; Newtonian viscosity increases with time but does not become as large of η_p for plastic-behavior period.

Table 1 (continued)

Culture	Shear rate[a] (s^{-1})	Viscometer	Flow curves given	Comments
S. griseus [53] (whole broth)	1–10	Brookfield	Yes	1. Pseudoplasticity exhibited throughout batch (see Fig. 16). 2. Apparent "Bingham plastic" behavior for $\dot{\gamma} > 5\ s^{-1}$ (see text). 3. "τ_y" and "η_p" for apparent "plastic" behavior both exhibit maxima (Fig. 17).
S. aureofaciens [54] (whole broth)	2–3	Brookfield and Contraves Rheomat	Yes	1. Initial viscosity high due to high starch concentration but decreases due to starch hydrolysis. Newtonian behavior early in batch.
S. noursei [52]	$< 0.2^b$	Brookfield (guard removed)	Yes	1. Newtonian behavior throughout batch despite mycelial character of organism. 2. Viscosity increase with fermentation time.
S. niveus [48] (whole broth)	in situ	No		1. "Power law" behavior throughout batch (see Sect. 2.3.2.5.).

[a] Shear rate was assumed equal to $10 \cdot N\ (s^{-1})$ when not given explicitly.
[b] Shear rate assumed as in [a], but differs greatly from values cited by authors. Error is suspected in reported values.

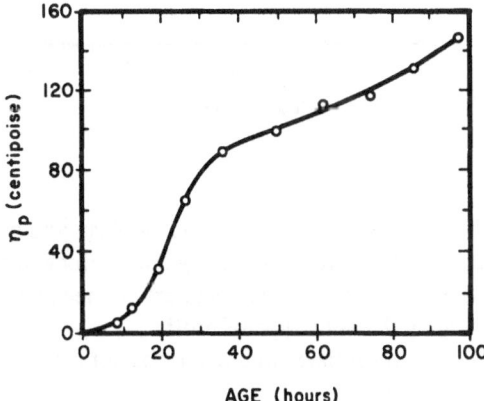

Fig. 13. Plastic viscosity (rigidity) vs. elapsed reaction time, P. chrysogenum [51]

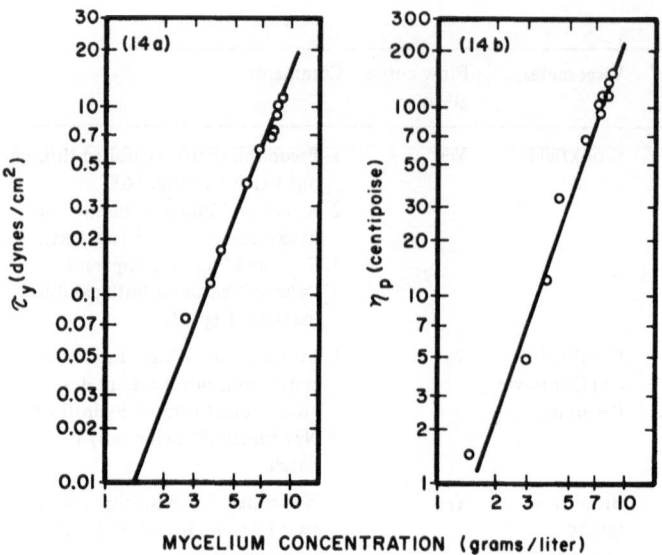

Fig. 14. (a) Yield stress and (b) rigidity *vs.* mycelium concentration time, *P. chrysogenum* [51]

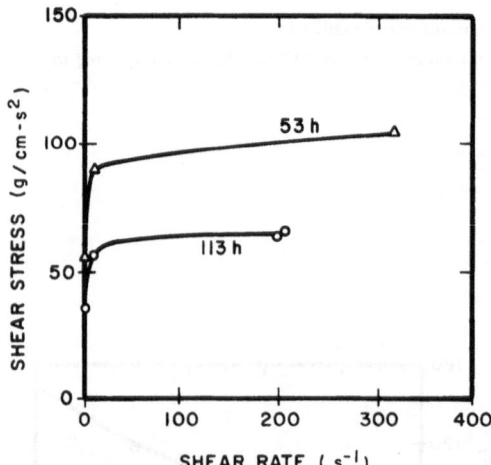

Fig. 15. Flow curves at two elapsed reaction times, *P. chrysogenum* [52]

(1) in all cases but one [18] morphology was not described
(2) rheological fundamentals were overlooked in many cases
(3) different types of viscometers and techniques were used by different investigators studying the same type of broth
(4) the appropriate viscometer was not always used—particularly for determination of yield stress

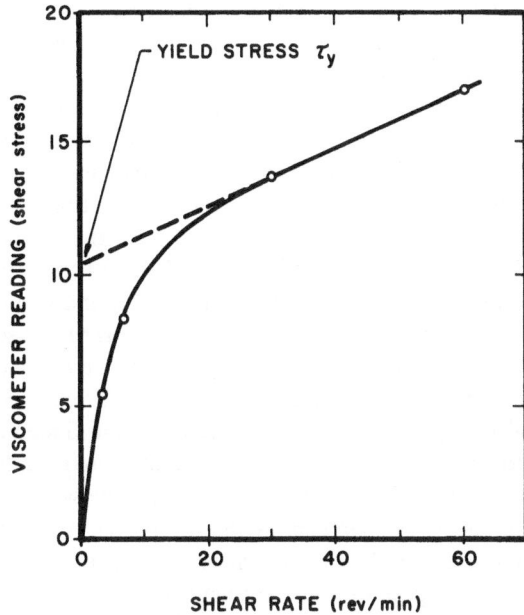

Fig. 16. Flow curve and "yield-stress" determination, *A. griseus* [53]

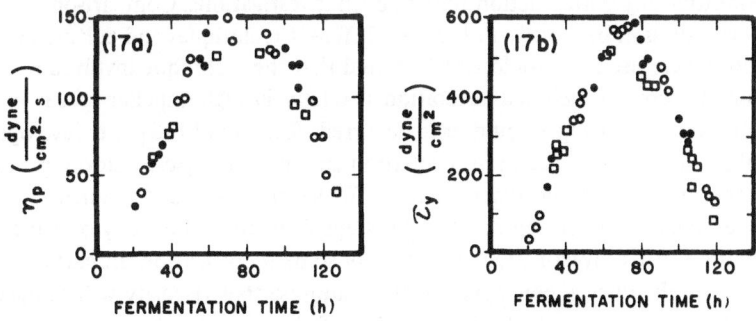

Fig. 17. (a) Plastic viscosity and (b) "yield-stress" *vs.* elapsed reaction time, *S. griseus* [53]

(5) instrument operating problems (e.g., phase separation at rotating bib, particle settling) appear to have been ignored in many cases.

Solomons and Weston [50] used consistency index (K) values for suspensions containing various concentrations of washed *A. niger* mycelia to correlate oxygen transfer measurements. However, they did not report the range of shear rates for which K's were determined and did not present experimental flow curves. Therefore, it is not possible to determine if their K values have any physical significance under the shear rate conditions of the oxygen transfer experiment. Furthermore, the oxygen transfer correlation itself does not take into account explicitly the variation of viscosity with impeller speed but is based only on the variation of K with mycelial concentration. Finally, they note that for the suspensions studied yield stress and K values are essentially equal. In the absence of flow curves and information regarding the shear rate range, it is difficult to judge the validity of this statement.

Deindoerfer and Gaden [51] reported Bingham plastic behavior for *P. chrysogenum* culture broths. As shown in Figs. 13 and 14 they found that τ_y and η_p both increase monotonically with reaction time and cell mass concentration. Deindoerfer and West [52] reported a much different behavior as can be seen from Fig. 15. Their results seem to suggest two distinct structure effects; one related to yield stress and the other to the sharp breaks in the flow curves. It is difficult to compare the report of Deindoerfer and West [52] with that of Deindoerfer and Gaden [51] because different techniques were used, neither of which was ideally suited for determination of yield stress. Furthermore, the media compositions were not the same in both investigations. Finally, there is some question concerning the shear rates reported by Deindoerfer and West [52]. They give viscosity data for shear rates of from 3 to 300 s^{-1}. However, the maximum rotational speed of the instrument they used is 60 RPM which corresponds to a shear rate of approximately 10 s^{-1} and they imply that their data was collected at much lower speeds (< 1 RPM).

Bongenaar *et al.* [42] and Roels *et al.* [18] used the "Turbine" viscometer to measure the rheological properties of *P. chrysogenum* broths and in both cases observed Casson fluid behavior. Yield stress increased monotonically and consistency index (K_c) decreased monotonically with reaction time in both investigations. Comparison with data cited previously is at best difficult because different techniques and culture medium compositions were used. It should also be noted that the technique involved in using the "Turbine" viscometer included deaeration at relatively high impeller speeds [42] and hence morphological changes could have occurred. Roels *et al.* [18] also developed a rheological model based on the Casson equation and morphological theories previously applied to the study of high polymer rheology. While the model and the method used to obtain experimental data do represent major steps forward in the study of broth rheology, they require further investigation and possibly some modification. In particular, the model is based on rheological concepts which presuppose that viscosity will be measured under conditions satisfying the basic definition of viscosity. As observed previously, the turbine does not satisfy these conditions. However, this is not to say that the incompatibility is so great as to make useless the overall approach. However, further work is required to define the range of applicability. Finally, it is interesting to note that Roels *et al.* [18] report that with only minor modification their model can be used to correlate the *A. niger* yield stress data of Solomons and Weston [50] and that in particular

$$\tau_y \, \alpha \, X^{2.5}, \tag{38}$$

where X is the mycelial concentration. This is somewhat surprising in that Solomons and Weston [50] reported K values and not yield stress values. While it is true that they stated the two are essentially equal for *A. niger* broths (actually washed-cell suspensions) studied, they provided no evidence. Since the shear rate range for which the K's were determined was not given, it is not possible to draw meaningful conclusions.

Deindoerfer and West [52] report that *S. griseus* broths exhibit Bingham plastic behavior early in the batch (24 h) but Newtonian behavior later (48–96 h); the Newtonian viscosity increases with time but never becomes as great as the original η_p. The effects of initial medium composition were not reported. Given what was probably a very low shear rate (previously discussed in connection with the work of these authors on *P. chrysogenum*) it is possible that the Newtonian viscosity reported was actually the zero-shear viscosity of the suspension. Richards [53], who also used a Brookfield but at shear rates of $1-10$ s^{-1}, reports pronounced pseudoplastic behavior throughout the culture although he prefers to interpret his results in a manner which implies Bingham plastic behavior. An experimentally determined flow curve given in Fig. 16 [53] illustrates his method for determining "τ_y" and "η_p" for this "Bingham plastic" behavior. Variations of "η_p" and "τ_y" with time are given in Fig. 17 [53]. The present author finds this practice objectionable because it gives a false impression of the rheological character of the fluid.

Tuffile and Pinho [54] report that for a culture involving *S. aureofaciens*, the initial medium exhibits pseudoplastic behavior (power law) due to a high starch concentration but that Newtonian behavior develops during the first 22 h as a result of starch hydrolysis. Beyond 22 h mycelial growth causes pseudoplasticity which can be expressed in terms of the power law equation. Typical plots of K and m values as functions of time reaction are given in Fig. 18 [54].

Deindoerfer and West [52] report that *S. noursei* cultures exhibit Newtonian behavior throughout the culture despite the mycelial character of the organism. This seems rather surprising and the author can only suggest that, as in previous discussion of the work of these authors, zero shear viscosities may have been measured.

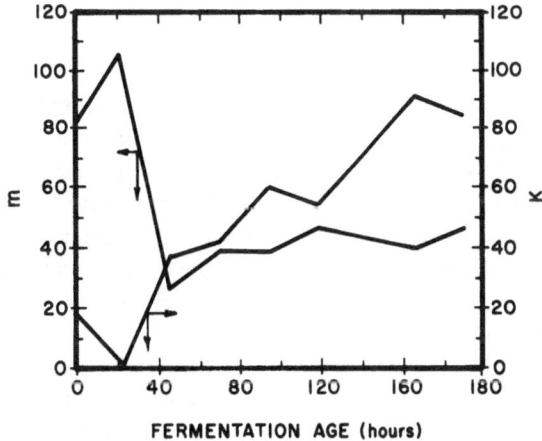

FERMENTATION AGE (hours)

Fig. 18. Variation of pseudoplastic characteristics with elapsed reaction time, *S. aureofaciens* [54]

3.2. Culture Fluids Containing Extracellular Polysaccharides

The rheological properties of a culture fluid containing an extracellular microbial polysaccharide is controlled by the concentration and molecular properties of the dissolved
polysaccharide. While the cells do have some influence, there is evidence that it is relatively minimal as illustrated in Fig. 19 for *Xanthomonas campestris* broths before and
after clarification [56]. Air bubbles may also effect rheological measurements. For
example, it is known that the yield value of beaten egg white results from the adsorp-

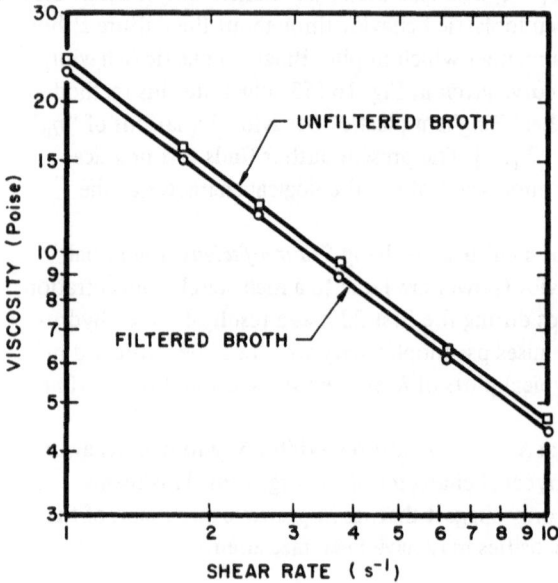

Fig. 19. Effect of cell mass on viscosity of culture fluid containing xanthan

tion of protein on bubble surfaces which promotes interparticle interference. However,
in the case of xanthan gum solutions, air bubbles seem to exert only minor influence
[56]. Nevertheless, the effects of solids and air bubbles in fluids containing polysaccharides other than xanthan do not seem to have been studied. Indeed, the problems have
not been mentioned in the literature being reviewed here. Therefore, no generalizations
should be made on the basis of the data reported in this paragraph. The possible effects
of air bubbles and solids content, particularly in the case of polysaccharide-containing
fluids, deserve greater consideration.
Solutions of high polymers usually exhibit complex non-Newtonian behavior and
extracellular microbial polysaccharides are no exceptions. For example, solutions of
xanthan gum and xanthan-containing culture fluids not only exhibit pronounced pseudoplasticity [20, 57, 58] as shown in Fig. 20 [57], but also yield stress [59] and thixotrophy [20]. Interestingly, the power law model provides an adequate description of

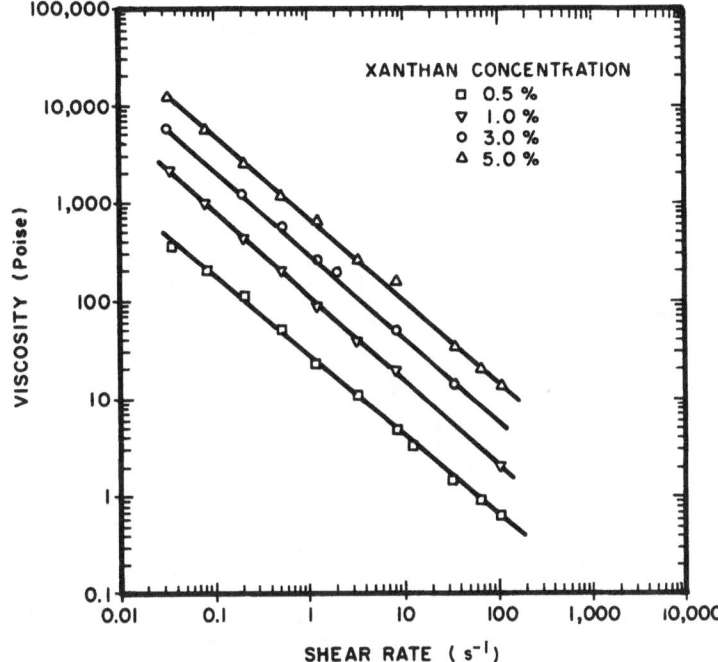

Fig. 20. Viscosity characteristics of culture fluids containing various concentrations of xanthan

pseudoplasticity over the shear rate range usually encountered in reactors [20, 57]. As illustrated in Fig. 21 [58] many other microbial polysaccharides exhibit pronounced pseudoplasticity, although power law behavior is not always observed. As a point of interest it should be noted that the method of plotting viscosity illustrated in Fig. 21 [58] was first introduced by Patton [60, 61] and can be quite useful.

LeDuy et al. [62] have performed the most extensive rheological study yet reported for polysaccharide-containing broths. They cultured *Pullularia pullulans* under various conditions in shake flasks and measured rheological properties as a function of both culture age and polysaccharide concentration. As shown in Fig. 22 [62] they report power law pseudoplastic behavior which increases to a maximum after several days and subsequently decreases. Although no substantial proof is given, they suggest that this behavior is associated with changes in average molecular weight and also suggest that the decline of apparent viscosity during the later stages of the fermentation result from the action of a depolymerase. Yield stress and thixotropy were observed but no details are given. No mention is made of any attempt to determine effects of cell mass and air bubbles on rheological properties.

Before going on it is worth noting that the apparent viscosity reported by LeDuy et al. [62] is that *measured* at a shear rate of $1.0 \, s^{-1}$. They refer to this as the *field apparent viscosity* (μ_{ap}). It is important to recognize that this is not necessarily equal to the consistency index (K) in the power law model as the latter is an extrapolated value which need not have a real physical significance at a shear rate of $1 \, s^{-1}$. Finally, LeDuy et al.

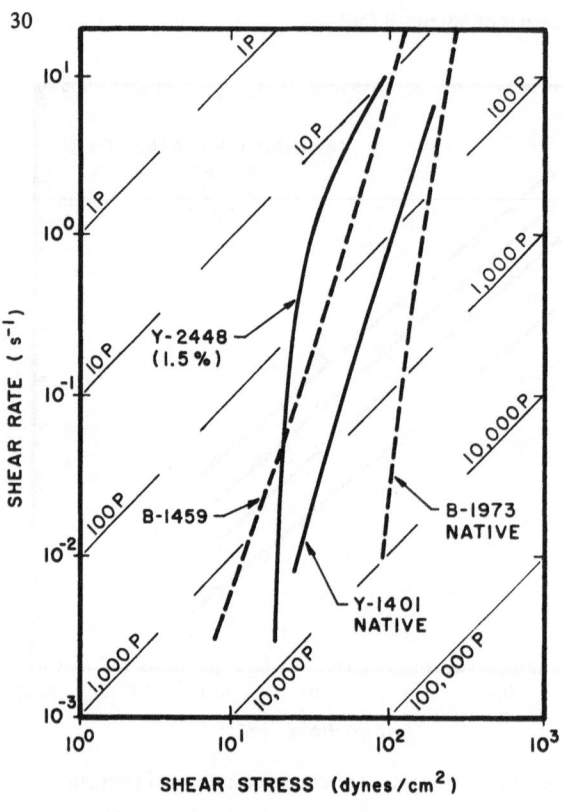

Fig. 21. Viscosity characteristics
of various non-Newtonian
microbial polysaccharides (in water)

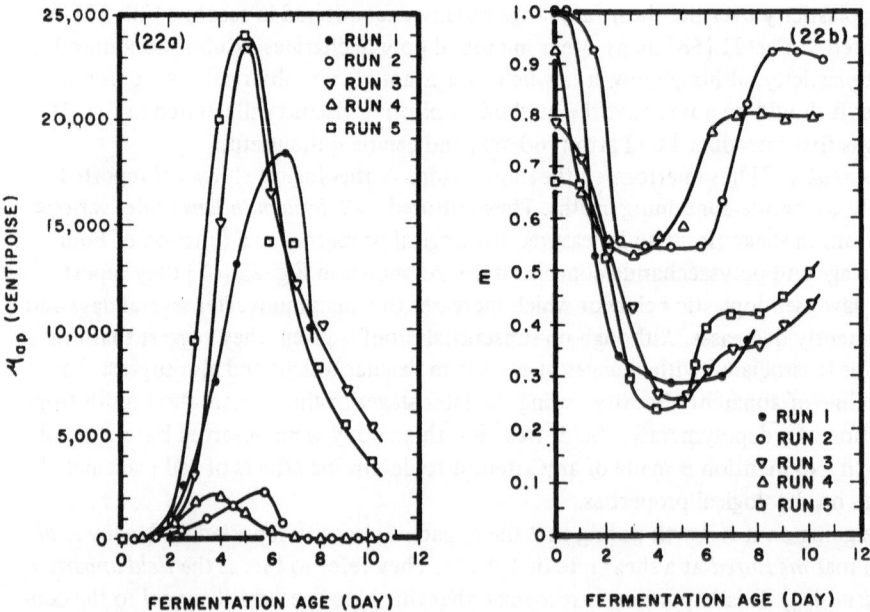

Fig. 22. Variation of (a) field apparent viscosity and (b) power law index with elapsed reaction time
for culture fluids containing pullulan, *Pullularia pullulans* (various initial conditions) [62]

contend that, "The viscosity of a non-Newtonian fermentation broth is more accurately represented by its field values". It is not clear to the author what is meant by "More accurately". If LeDuy *et al.* mean viscosity is measured more accurately at 1 s^{-1} by μ_{ap} than by the extrapolated value of K, the author must agree. However, if they mean μ_{ap} is a more accurate measure in general, the author must disagree since he feels that such a statement has no meaning if the shear rate range is not specified.

The rheological properties of *X. campestris* culture broths were studies in the author's laboratory [57] and as noted previously (Fig. 20), power law behavior was observed throughout. Cell mass concentration, which reached a maximum value of 0.6%, had little or no effect on viscosity measurements [56, 57]. It was also found that while K increased continuously with xanthan concentration (and hence reaction time), m decreased to ~ 0.15 as the concentration rose to 0.5% and then remained constant (final concentration $\cong$ 3.0% xanthan). The medium used in these experiments was primarily lactase-hydrolyzed ultrafiltered cottage cheese whey [62, 63]. Viscoelastic effects (e.g., a pronounced Weissenberg effect) [20, 57] were observed also, but unfortunately these were not studied quantitatively. Others [63, 66] have reported viscosity histories of xanthan synthesis. In all these cases, monotonic increase of viscosity with reaction time is reported, but basic rheological behavior is not discussed. Similar reports are available for other microbial polysaccharide-containing culture broths [67, 70]. Published rheological properties of culture fluids containing microbial polysaccharides are summarized in Table 2.

Only brief mention has been made regarding rheological characteristics other than shear-dependent viscosity. However, the extensive body of rheological data for general polymer solutions indicates that we should anticipate that culture broths containing microbial polysaccharides will exhibit effects such as yield stress, thixotropy, and viscoelasticity. This is important since such effects can be closely related to molecular configuration, molecular weight, and molecular weight distribution. Therefore, they may be valuable measures for monitoring, process control, and quality control. Furthermore, they can influence strongly mixing, mass transfer, and heat transfer during the reaction. The definitely warrant further attention.

Finally, the available evidence suggests that standard rotational viscometers are suitable for studying broths containing polysaccharides as the air bubbles and solids present do not appear to interfere significantly with viscosity measurements. Nevertheless, further verification of this contention would be prudent. It would also be worthwhile to compare results obtained using standard viscometers with those obtained using the methods proposed by Bongenaar *et al.* [42] and by Wang and Fewkes [48].

Table 2. Rheological properties of culture fluids containing extracellular microbial polysaccharides

Culture	Shear rate[a] (s^{-1})	Viscometer	Flow curves given	Comments
Aureobasitium pullulans [62]	10.2–1020	Fann V-G (model 35)	Yes	1. Power law behavior throughout batch. 2. Both m and K vary with reaction; K exhibits a maximum, m a minimum (Fig. 22). 3. Variation in rheological conditions of batch. 4. Some evidence for depolymerase activity near end of batch.
Xanthomonas campestris [57]	0.0035–100	Weissenberg Rheogoniometer (cone and plate)	Yes	1. Power law behavior; K continually increases but m levels off to a constant value after concentration of xanthan becomes greater than 0.5% (Fig. 22). 2. Cell mass (0.6% max) had little or no effect on viscosity measurements.
X. campestris [56] (whole broth and filtered broth)	1.0–10.0	Brookfield (guard removed)	Yes	1. Whole and filtered broth obeyed power law (Fig. 19). 2. Cells had minor effect on K, virtually no effect on m.
X. campestris [63, 66] (whole broth)	5 or 10	Brookfield (guard removed)	No	1. Viscosity increases monotonically with reaction time. 2. Nature of rheological behavior not discussed.
Aureobasitium pullulans [67] (whole broth)	0.17	Brookfield	No	1. Viscosity increases monotonically with reaction time. 2. Nature of rheological behavior not given.
Penicillium funiculosum [68]	not given	Brookfield	No	1. Medium contains dextran. 2. Viscosity measurement provides excellent measure of dextranase production. 3. Nature of rheological behavior not discussed.
Hansenula holstii [69]	5.0	Brookfield	No	1. Viscosity increases monotonically with reaction time. 2. Nature of rheological properties not discussed.
Rhinocladiella mansonii [70]	5.0 1 RPM (γ not given)	Brookfield; Brookfield-Wells cone and plate	No.	1. Viscosity shows extreme dependence on concentration; gentle increase for concentration up to 0.3% then drastic rise.

[a] Shear assumed equal to $10 \cdot N$ (s^{-1}) when not given explicitly.

3.3. Yeast and Bacterial Cultures

Surprisingly little experimental data have been published for cultures composed of cells which do not exhibit mycelial growth. Indeed, the only published data is for yeast cultures and illustrates only the effect of concentration on viscosity. Deindoerfer and West [71] report that the original yeast (*S. cerevisiae*) suspension (water)viscosity data of Eirich [72] can be correlated well by the Vand equation:

$$\eta = \eta_s (1 + 2.5 \, \phi + 7.25 \, \phi^2), \tag{39}$$

where η_s is the viscosity of the solvent and ϕ is the volume fraction of cells. No mention was made of the instrument used or of the shear rate. Shimmons *et al.* [73] used a Contraves DDA-005 in-line rotational viscometer operated at 2200 s^{-1} to measure the viscosity of water suspensions of *S. cerevisiae*. They report that

$$\eta = \frac{2.02}{1 - \phi} \, \eta_s + 1.36 \, \phi. \tag{40}$$

There is substantial disagreement between Eqs. (39) and (40) at volume fractions as small as 0.05 and the disagreement increases markedly with increasing ϕ. The absence of information regarding complete experimental details makes it impossible to formulate an explanation.

Viscosity-concentration data for yeast concentrates (washed?) reported by Modéer [74] is given in Fig. 23. While it is unlikely that cell mass concentrations during growth will

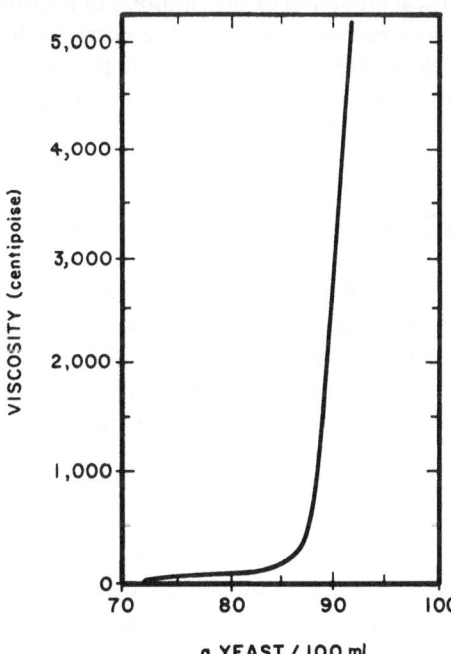

Fig. 23. Viscosities of concentrated yeast suspensions [74]

ever fall in the range considered, it is quite likely that it will do so during recovery. Therefore, the recovery system and its controls must be designed to cope with the drastic rise in viscosity at concentrations above 87 g packed yeast/100 ml concentrate. Unfortunately, Modéer has given no information regarding the viscometer he used or the shear rate or rates he considered. A complete introduction to the theory of the rheological properties of suspensions is given by Goldsmith and Mason [75].

4. Mixing

4.1. Introduction

In essence, the reactor's raison d'être is to provide a controlled environment which promotes, in the most economic fashion possible, the transport of mass and heat at rates such that the intrinsic characteristics of the microbial population determine the process kinetics.

The observed rate of reaction for a particular organism is dependent on both the nature of the physico-chemical environment and the hydrodynamic state in the immediate vicinity of the organism. In the general case we must expect that both of these will vary both temporally and spatially. The extents of these variations depend not only on the characteristics of the organism but also on mixing conditions which, for a given reactor configuration and fixed operating conditions, depends strongly on the rheological properties of the culture. For a stirred-tank reactor, high viscosity and non-Newtonian behavior usually lead to decreased homogeneity (e.g., formation of stagnant regions) which, in turn, results in poor performance (e.g., diminished yield and productivity): Complete homogeneity is the preferred mixing state for the stirred-tank fermentor. (This is not necessarily true for other reactor types such as the plug flow reactor.)

4.2. Mixing Time

To establish the desired state of homogeneity in a stirred-tank reactor the agitator must provide good macromixing (bulk circulation or convection flow) [76, 78]. Good micromixing (mixing at or near molecular scale) [76, 78] is also desirable but usually not as important as good macromixing [76, 78]. One frequently used measure of agitator "homogenizing" ability is the *mixing time,* θ_M, which is generally defined as the time required to reach a specified degree of uniformity after a tracer pulse (e.g., acid or base, concentrated salt solution, heated fluid) has been added to the previously homogeneous agitated mass. The experimental procedure involves continuous monitoring of the characteristic affected by the tracer (e.g., pH) and therefore provides, in addition to the mixing time, some insight to the nature of the mixing process as can be seen from typical results given in Fig. 24 [79]. In general, for a given fluid the nature of the mixing process (and hence the mixing time) depends on the overall reactor geometry, the impeller

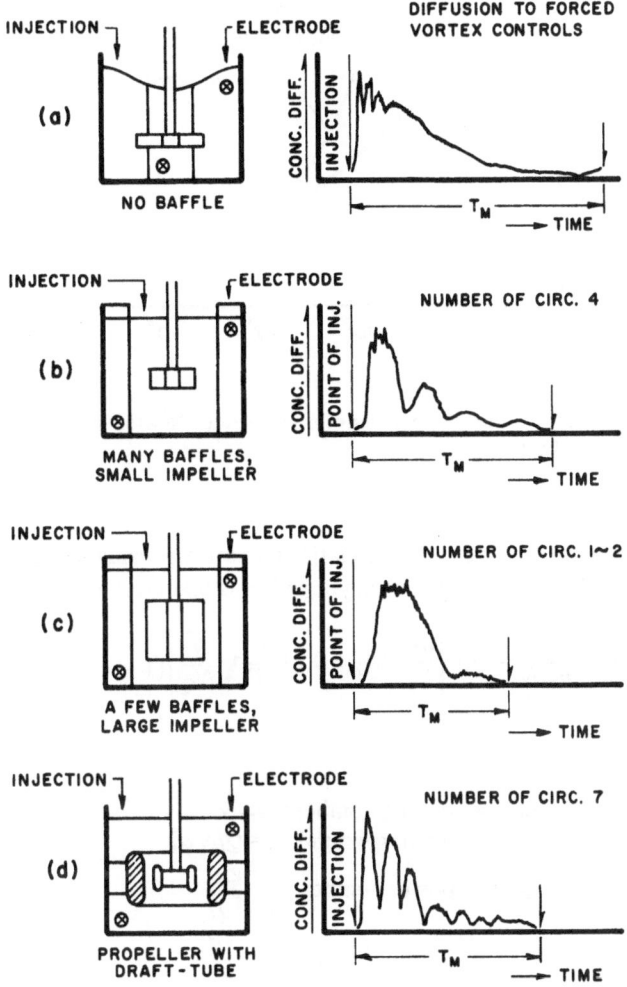

Fig. 24. Mixing time characteristics of various impellers [79]

type [80–83], and the impeller speed. For a given reactor and fixed operating conditions, mixing time can be affected strongly by the rheological nature of the fluid as can be seen from Fig. 25 [81] in which mixing time correlations for Newtonian and non-Newtonian fluids are compared (turbine impeller).

4.2.1. Newtonian Fluids

Several mixing time correlations have been developed for Newtonian fluids [80–88]. However, due to the complexity of the mixing process, the fact that different tracer methods give different results, and the somewhat arbitrary definition of the mixing time, different correlations can give significantly different estimates of mixing time.

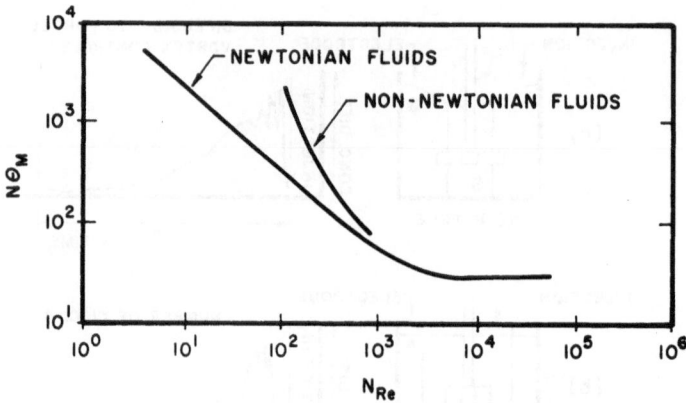

Fig. 25. Mixing time correlations for non-aerated Newtonian and non-Newtonian fluids, turbine impeller [81]

Nevertheless, available correlations do serve as valuable guides. One of the more recent correlations for top-entering impellers is that given by Nagata [88]:

$$\frac{1}{N_m} = \frac{1}{\theta_m N} = 0.1 \left\{ \frac{d_i}{T} \, {}^3N_Q + 0.21 \, \frac{d_i}{T} \sqrt{\frac{N_P}{N_Q}} \right\} (1 - e^{-13(d_i/T)^2}) \, (N_{Re} < 10^4), \qquad (41)$$

where

$$
\begin{aligned}
N_m &= \text{mixing number} = \theta_m N & (42)\\
N_Q &= \text{discharge number} = Q/N \, d_i^3 & (43)\\
N_P &= \text{power number} = \frac{P g_c}{\rho N^3 d_i^5} & (44)\\
N_{Re} &= \text{Reynolds number} = \frac{N d_i^2}{\nu} &
\end{aligned}
$$

and

$$
\begin{aligned}
\theta_m &= \text{mixing time}\\
N &= \text{impeller rotational speed}\\
d_i &= \text{impeller diameter}\\
T &= \text{tank diameter}\\
Q &= \text{volumetric circulation rate in tank}\\
P &= \text{power input}\\
\rho &= \text{fluid density}\\
\nu &= \text{kinematic viscosity.}
\end{aligned}
$$

Values of N_P and N_Q for use in Eq. (41) vary with impeller type and vessel geometry and can be obtained from standard correlations [24–26, 89, 90]. Nagata [91] has also shown that one can obtain a rough approximation of mixing time from

$$N\theta_m = \frac{3V}{N_Q \cdot d_i^3},$$ (45)

where V is the liquid volume. For other correlations the reader is referred to the literature [80–89], the reviews of Wang and Humphrey [92] and Hyman [93] and in particular to the comprehensive treatise of Nagata. However, in all cases the reader should be aware that these correlations for Newtonian fluids were developed for unsparged systems containing no cell mass and may not be applicable to aerated reactors (see below).

4.2.2. Mycelial Cultures

The early reports of Phillips and Johnson [94], Maxon [95], and Steel and Maxon [96, 97] demonstrate clearly that good bulk mixing of mycelial cultures is important but difficult to achieve and lead one to the conclusion that it would be quite valuable to correlate a quantitative measure of mixing quality (e.g., mixing time) with rheological properties and impeller characteristics. However, the author is aware of only two reported works addressed to the development of such correlations. In both cases standard vessels and turbine impellers were employed.
Blakebrough and Sambamurthy [98] found that mixing times for a simulated non-Newtonian mycelial broth (1.6% paper pulp suspension-rheological properties not reported) can be correlated with the *momentum factor* which they defined as

$$Mf = Nd_i x NWL\,(d_i - W)$$ (46)

where

L = impeller blade height
W = impeller blade width.

Their correlation is presented in Fig. 26 [98]. The effects of impeller design and aeration are clearly demonstrated. The effect of aeration is particularly noteworthy because the

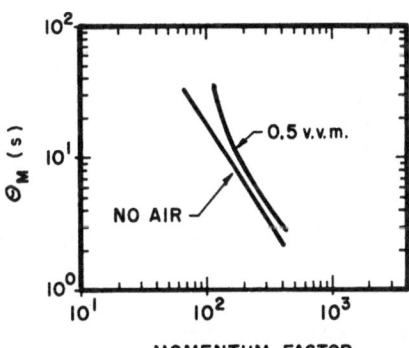

Fig. 26. Mixing time correlations for aerated and non-aerated simulated culture broths, turbine impeller [98]

other available mixing time correlations are based on experiments involving non-gassed fluids. The work of Blakebrough and Sambamurthy suggests that these other correlations should be modified for bio-applications. Finally, it is important to note that the correlation does not include rheological properties explicitly and hence its range of applicability can not be predicted easily.

Wang and Fewkes [99] report that mixing times for actual broths containing *Streptomyces niveaus* can be correlated by the method of Norwood and Metzner [84] in which a dimensionless number, N_m^1, defined as

$$N_m^1 = \frac{\theta_m (N d_i^2)^{2/3} g^{1/6} d_i^{1/2}}{y^{1/2} T^{3/2}}, \tag{47}$$

where g is the acceleration due to gravity and y is the liquid depth, is correlated with the impeller Reynolds number given by

$$N_{Re} = \frac{N d_i^2 \rho}{\eta_a}. \tag{48}$$

This correlation was originally developed for non-gassed Newtonian fluids but it has been suggested [84, 89] that it may be used for non-gassed non-Newtonian fluids at relatively high Reynolds numbers if a modified, non-Newtonian Reynolds number is used [27, 84]. Wang and Fewkes chose to use the modified Reynolds number suggested by Calderbank and Moo-Young [25, 26] As can be seen from Fig. 27 [99] their results

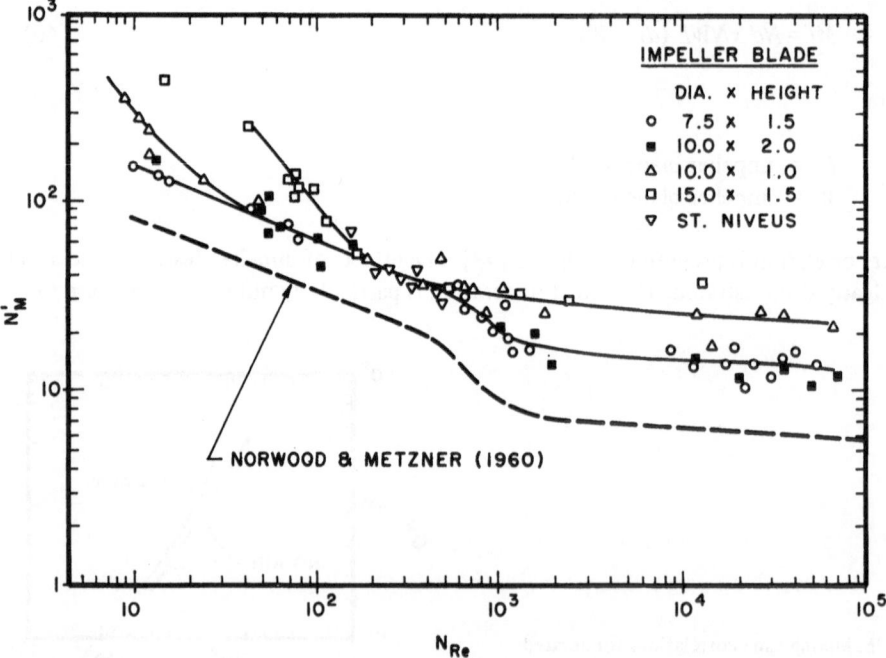

Fig. 27. Mixing time correlations for *S. niveus* culture fluid for various turbine impeller configurations [99]

are in good agreement qualitatively with the original Norwood-Metzner [84] correlation although they differ quantitatively. The discrepancies might be due to aeration effects as demonstrated by the work of Blakebrough and Sambamurthy [98]: Wang and Fewkes [99] did not specify whether mixing times were determined under aerated or non-aerated conditions nor did they mention the extent of gas hold up. Another source of the disagreement might be inappropriate use or measurement of rheological properties; it might be recalled (Sect. 2.4.2.5.) that there is some question regarding the validity of *in situ* viscosity measurements made by these authors. One must also recognize that the definition of mixing time is somewhat arbitrary and its evaluation is partially subjective. Also, it is possible that the Norwood-Metzner correlation is not valid here (assuming the experimental conditions were as specified by Norwood and Metzner). Finally, the reader is cautioned about what appears to be a typographical error in Wang and Fewkes' [99] equation for the Norwood-Metzner mixing number. Wang and Fewkes [99] also suggest that mixing time is a useful parameter in correlating oxygen transfer rates for various turbine impellers over a range of cell mass concentrations (and hence a range of rheological properties) and recommend that it be considered as an important factor in scale-up calculations. However, they do not recommend scale-up on the basis of constant mixing time. In the same paper they also propose a relationship between mixing characteristics and a quantity they define as the *pseudo-critical dissolved oxygen concentration* (C_c^*) which they visualize, "... as the minimum bulk dissolved oxygen concentration which is needed to effect a positive change in the cells' ability to take up oxygen". They demonstrate that C_c^* can be correlated very well with the *shear to flow ratio* which is defined as

$$\frac{\text{Impeller shear}}{\text{Impeller flow}} \; \alpha \; \frac{N^2 d_i^2}{N d_i^3} \; \alpha \; \frac{N}{d_i} \; (\text{cm}-\text{s}^{-1}). \qquad (49)$$

The author believes that despite the discrepancies noted above the work of Wang and Fewkes [99] is a valuable contribution to the state of the art and should be used as a basis for further and much needed work related to mixing effects in non-Newtonian culture fluids.

4.2.3. Culture Fluids Containing Microbial Polysaccharides

No mixing time correlations have been reported for culture fluids containing microbial polysaccharides. However, some initial experiments have been conducted in the author's laboratory to determine the effects of gum concentration, agitator speed, and aeration levels on mixing times for broths containing xanthan [100]. Some typical pH traces (response to pulse addition of acid or base) are presented in Fig. 28. This data was obtained using a 7-l partially baffled vessel the contents of which were agitated by three 6-bladed (45° pitch) disk turbines spaced one impeller diameter apart. The vessel diameter was 14.1 cm, the tank diameter to impeller diameter ratio was 1.8, and the vessel contained 5 liters of culture broth (including cells at a concentration of approximately 0.6%). While these results are not suitable for the development of a general mixing time correlation, they do demonstrate the profound effects of the polysaccha-

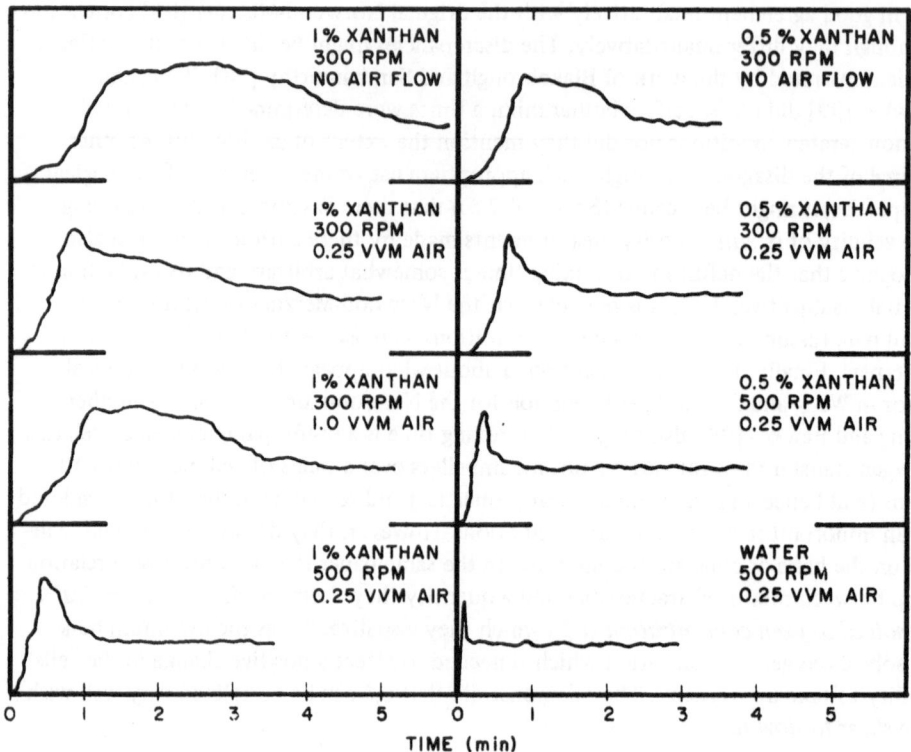

Fig. 28. Mixing times for culture broths containing xanthan under various conditions of agitation and aeration

ride rheological properties on mixing. These effects are more pronounced than those observed with fungal broths and appear to be more sensitive to operating conditions (e.g., agitator speed, air flow rate). However, much more work is required before we can make quantitative generalizations and meaningful comparisons.

4.3. Impeller Design

Further studies of mixing in bioreactors should include not only the development of reliable correlations but also the design of new impellers better suited than turbines to mixing non-Newtonian culture fluids. Only Steel and Maxon [97] seem to have given this serious consideration. They demonstrated clearly the advantages of the multiple rod impeller over turbine impellers in providing homogeneous oxygen transfer conditions in fungal culture broths. While not concerned with growth, other investigators have shown that turbine impellers are not particularly efficient for mixing viscous non-Newtonian fluids [80–83] (see Leamy [101] for opposite opinion). The reason for this in the case of pseudoplastic fluids is clear from Norwood and Metzner's [81] description of pseudoplastic flow in the immediate vicinity of a turbine. Quite close to the impeller the shear rate is high and hence the viscosity is low; fluid pumping conditions are good as are mass transfer conditions. However, the shear rate drops rapidly with distance from the impeller and hence the viscosity increases rapidly. Fluid motion is therefore damped and mass transfer rates are diminished. The overall result is poor bulk circulation and a limited region of good mass transfer.

It has been observed also that in turbine-agitated aerated mycelial cultures, most of the gas flow is in the vicinity of the impeller [38]. This results in (1) greater gas channeling and lower residence time, (2) increased bubble size (3) the deterioration of bulk mixing and (4) poor distribution of oxygen throughout the vessel. All of these factors contribute to the greatly diminished oxygen transfer rates caused by the mycelia [51]. Culture fluids containing polysaccharide gums seem to behave the same way [102]. This behavior, which is difficult to describe and to predict quantitatively, also complicates the task of developing reliable power consumption correlations (Sect. 4.4.).

In seeking methods to overcome this problem one must recognize that the rheological phenomena causing maldistribution of the gas are not identical for both types of culture fluids. In the case of mycelial broths, there tends to be a mycelia-free region in the immediate vicinity of the impeller (similar to behavior of such fluids in rotational viscometers as discussed in Sect. 2.4.2.4. and roughly analogous to the "tube pinch" effect [39] in capillary flow) beyond which there is a region of rapidly increasing viscosity as described above. For broths containing polysaccharides there is only the region of increasing viscosity. Therefore, a different impeller design will probably be required for each rheological type. In any event, greater attention must be paid to proper impeller design if we are to overcome problems of poor and unpredictable mixing and to establish more rational and reliable design methods.

Bulk mixing and mass transfer can be improved by employing larger turbines and by increasing impeller speed but generally both measures are required and hence the increase in power consumption is considerable. Therefore, impellers [80–83] (e.g., helical ribbon) which have been found to be more efficient than the turbine for mixing viscous non-Newtonian fluids should be investigated for fermentation applications. However, it should be borne in mind that existing information for these more efficient impellers is based on studies of non-aerated systems. Studies of reactor applications must include investigation of the effects of aeration and solids concentration. Also, it is important that mixing quality, mass transfer rates, and power consumption be investigated simultaneously whenever possible.

Finally, any future study of non-Newtonian mixing and mass transfer should include as complete a rheological characterization of the fluid(s) as possible. Certainly, the dependence of viscosity on shear rate should be determined as a routine matter, but other properties such as yield stress, thixotropy and viscoelasticity warrant consideration when possible. In this regard it is interesting to note that Chavan *et al.* [103] report that, "... mixing times in viscoelastic fluids will be about twice as long as in inelastic liquids of comparable consistency".

4.4. Power Consumption

Correlations for power consumption by non-gassed non-Newtonian fluids have been
well established for a variety of single impellers and are reviewed adequately elsewhere
[24–27, 92, 93, 104–107].
A correlation [108] for multiple turbine impellers has been developed for non-gassed
non-Newtonian fluids but its range of applicability has not been well established nor
have the effects of impeller spacing been elucidated to any great extent [109].
Power requirements for aerated mycelial culture fluids have not been correlated ade-
quatly despite the considerable attention given this subject which has been discussed in
a number of excellent reviews [53, 92, 105, 109].
Power requirements for mixing aerated non-Newtonian solutions of polymers have
received very little attention in general and aerated culture fluids containing microbial
polysaccharides appear to have been considered in only one reported investigation.
Charles *et al.* [20] found that for xanthan solutions having concentrations ranging from
0.5% to 5.0%, aerated at from 0.25 to 1 vvm, and agitated at speeds up to 500 RPM by
3 standard turbine impellers (T/d_i = 2.54) in a fully baffled 270 L reactor (200 L working volume), power consumptions could not be even roughly approximated by existing
correlations. Clearly, far more work is required to develop reliable power consumption
correlations for both mycelial cultures and those containing microbial polysaccharides.
(In contrast to this opinion, Leamy [101] suggests that available correlations are adequate.) Such work should include measurement of rheological properties and mixing
times (or some other measure of mixing effectiveness).

5. Mass Transfer

5.1. Mass Transfer and Heterogeneous Catalysis; Basic Concepts

5.1.1. Two Phase Systems

Microbial growth is a heterogeneous catalytic reaction: The catalyst and substrates
(reactants) are located in different phases and hence substrates must diffuse across at
least one diffusion barrier ("boundary layer") in order to reach the reactive cell surface.
(The characteristics of such diffusion boundary layers are discussed at length in various
treatises concerned with mass transfer and transport phenomena [110–113].) The
simplest case is depicted schematically in Fig. 29. The surface reaction occurs at a rate
dependent on the substrate concentration and environmental conditions at the surface.
[Pore diffusion effects will not be considered here. They are known to be important
when cells grow in aggregates such as in pellet-like growth of fungi but may also be
important in unicellular growth depending on the nature of (1) the cell wall, (2) any
capsule surrounding the cell and (3) invaginations of the cell surface.] The difference
between surface conditions and those in the bulk substrate phase depends on the resis-

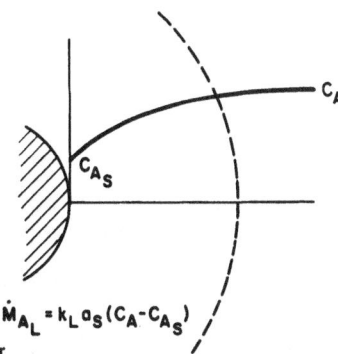

Fig. 29. Diffusion model for two phase interphase mass transfer

tance offered by the diffusion barrier and this is influenced strongly by mixing conditions and rheological properties.

As a simple example, consider a case in which conditions at the cell surface are such that the Monod equation may be approximated by the first order expression

$$\frac{dC_{AS}}{dt} = -k_1 C_{AS}, \tag{50}$$

where C_{AS} is the substrate concentration at the surface and k_1 is the reaction rate constant. The rate, $\dot{M}_{A_L}$, at which substrate can move from the substrate phase to the cell surface may be expressed as

$$\dot{M}_{A_L} = k_L a_S (C_A - C_{AS}), \tag{51}$$

where C_A is the substrate concentration in the substrate phase and a_S is the surface area of the cell per unit volume of reactor. The term k_L is called the *mass transfer coefficient*. Its value depends on the physicochemical properties of the substrate in the given medium and on the intensity and nature of culture mixing which, in turn, depend in an interactive way on the rheological properties of the fluid phase. Therefore, quantitative correlations for k_L tend to be complex even for systems containing only two phases and a given correlation usually is applicable to only a rather limited set of conditions. Furthermore, while a number of such correlations [114–118] have been developed for stirred-tank reactors, none have been based explicitly on process data and all are essentially unproven for analysis and design of fermenters. Nevertheless, as a rule of thumb one may assume generally that the mass transfer coefficient will increase (towards some limiting value) as the intensity of mixing increases: Turbulent conditions tend to enhance mass transfer rates.

In the general case the observed rate of reaction will depend on both the intrinsic rate of reaction and the rate of mass transfer. To a good approximation one can say that the rate of reaction is equal to the rate of mass transfer:

$$k_1 C_{AS} = k_L a_S (C_A - C_{AS}). \tag{52}$$

Eq. (52) may be combined with Eq. (50) to give

$$-\frac{dC_{AS}}{dt} = r_A = \frac{k_1 C_A}{1+\dfrac{k_1}{k_L a_S}}, \tag{53}$$

where r is the *observed* rate of reaction (*overall* or *global* rate of reaction). This result makes evident the fact that if the inherent rate of reaction is much more rapid than the mass transfer rate $\dfrac{k_1}{k_L a_S} \gg 1$ the observed rate of reaction is governed by the mass transfer rate:

$$r_A \doteq k_L a_S C_A \tag{54}$$

However, if mass transfer is much more rapid than the reaction rate $\dfrac{k_1}{k_L a_S} \ll 1$, the observed rate is dependent on the intrinsic kinetics of the reaction:

$$r_A \doteq k_1 C_A. \tag{55}$$

It is generally preferable to operate under conditions favoring reaction rate control. This requires maximization of the mass transfer coefficient, k_L, and hence consideration of rheological properties.

The effects of a_S on the mass transfer rate must also be considered but these are difficult to analyze because morphological properties influence rheological properties and also introduce questions requiring consideration of "pore" diffusion. Space does not permit a discussion of this topic and hence the reader is referred to other reviews [119–122] for background information and further references. However, we should note here that the complex interactions between a_S (or morphology), rheological properties, and mass transfer coefficients have received almost no serious consideration at either the experimental or the theoretical level. The work of Roels et al. [18] is one of the few exceptions to this statement and could provide the basis for future investigation (experimental and theoretical) in this area.

5.1.2. Three Phase Systems

A very important case not covered completely by the preceeding treatment is that in which the substrate (e.g., oxygen) must diffuse first into the bulk liquid phase and then to the cell surface. This case must be analyzed separately because in addition to the fact that a new diffusional resistance (see below) must be considered, the presence of the new phase (e.g., gas) effects mixing characteristics and hence mass transfer from the bulk liquid to the cell surface.

The mass transfer path for a gaseous component (immiscible liquids and solids may be treated similarly) is shown in simplified form in Fig. 30. The rate of mass transfer, $\dot{M}_{AG}$, from the gas phase to the liquid phase may be expressed as

$$\dot{M}_{AG} = K_L a_G (C_A^* - C_A), \tag{56}$$

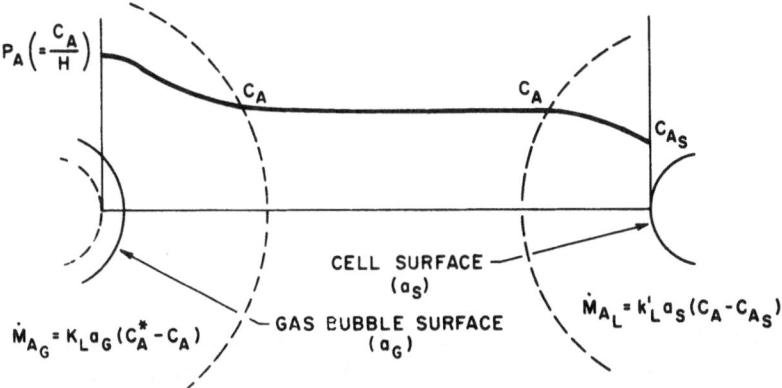

Fig. 30. Diffusion model for three phase interphase mass transfer

where K_L is the mass transfer coefficient for this interphase transport process, a_G is the surface area of the gas phase (bubbles), C_A^* is the maximum concentration of the solute (the concentration of A when the liquid is saturated with A at the prevailing partial pressure of A in the gas phase), and C_A is the actual concentration of A in the liquid phase. To a good approximation we can say

$$\dot{M}_{A_G} = \dot{M}_{A_L} = k_1 C_{A_S}. \tag{57}$$

Eqs. (50), (51), (56), and (57) may be combined to give

$$r_A = \frac{k_1 C_A^*}{1 + \dfrac{k_1}{K_L a_G} + \dfrac{k_1}{k_L^1 a_S}} \tag{58}$$

[Note that k_L in Eq. (51) has been replaced by k_L^1 to emphasize that this mass transfer coefficient may change when gas bubbles are present.]
If both rates of diffusion are very rapid in comparison to the intrinsic rate of reaction, $\dfrac{k_1}{K_L a_G} \ll 1$ and $\dfrac{k_1}{k_L^1 a_S} \ll 1$, the global reaction rate, r, is determined by the rate of the chemical reaction:

$$r_A = k_1 C_A^*. \tag{59}$$

In this case the reaction proceeds at the maximum rate allowed by the gas phase partial pressure of A. (C_A^* is related to the gas phase partial pressure by Henry's law; $C_A^* = H \cdot P_A$ where P_A is the partial pressure of A and H is the Henry's law "constant".) In a typical aerobic fermentation the observed rate is kinetically determined during the early stages while the cell mass concentration is low. However, at some point the cell mass concentration becomes so large that the rate at which oxygen can be supplied determines the observed rate. Under such circumstances, oxygen is consumed as rapidly as it approaches the cell surface and hence $C_{A_S} = 0$.

The observed rate is then

$$r = k_L{}^1 a_S C_A = K_L a_G (C_A^* - C_A) \tag{60}$$

and therefore

$$r = \frac{C_A^*}{\dfrac{1}{K_L a_G} + \dfrac{1}{k_L^1 a_S}}. \tag{61}$$

In general, then, both mass transfer resistances will influence the observed reaction rate, r. In the past, it usually has been assumed that the limiting resistance is that associated with transfer of oxygen from the gas phase to the liquid:

$$r = K_L a_G C_A^*. \tag{62}$$

However, the author [123] and others [99] have found that this may not be so for non-Newtonian systems such as those containing fungi or microbial polysaccharides and that further work is required to identify the limiting mass transfer resistance (s) in these cases. In this regard it is important to note that the controlling mass transfer resistance will be determined not only by the ratio of k_L^1 and K_L but also by the ratio a_S/a_g which depends on morphological characteristics and mixing conditions. Further complexity is introduced by the fact that mixing conditions are dependent on rheological properties which may themselves be dependent on culture morphology. Finally, it is worth re-emphasizing that the various mass transfer coefficients depend in large measure on overall mixing conditions; anything that alters the state of mixing can affect the values of the mass transfer coefficients. For example, it is quite likely that under the same operating conditions k_L will not be the same for aerated and non-aerated cultures. Furthermore, mixing states can vary dramatically with scale—particularly for non-Newtonian fluids. Thus, one reason that available mass transfer correlations for culture fluids are too often scale dependent is that they usually do not include explicit measures of mixing effectiveness such as the mixing time, θ_M. Exceptions to this are represented by the work of Blakebrough and Sambamurthy [98] and of Wang and Fewkes [99]. The author believes that those contemplating further work in this area should give careful consideration to these two contributions and should attempt to incorporate into their own correlations both an explicit measure of mixing quality (although this measure need not necessarily be θ_M) and rheological properties.

5.2. Oxygen Transfer

5.2.1. Mycelial Cultures

Oxygen transfer to mycelial culture fluids has been studied widely but little progress has been made toward the development of correlations useful for design and scale-up because, with the exception of the recent work of Wang und Fewkes [99] (Sect. 4.2.2.), rheological properties and mixing parameters have not been included explicitly in published correlations. Chain et al. [124], Phillips and Johnson [94], Maxon [95], and Steel and Maxon [96, 97] demonstrated the importance of mixing characteristics but ignored (quantitatively) rheological properties and did not quantify mixing parameters. Deindoerfer and Gaden [51], Solomons and Weston [50] and Tuffile and Pinho [54] studied the effects of rheological properties but ignored mixing behavior. [There is also some question regarding the accuracy and significance of viscosity measurements made by these investigators because in all cases they used standard rotational viscometers (see Section 2.4.2.4.).] Blakebrough and Sambamurthy [98] studied oxygen transfer to simulated fungal broths (1.6% paper pulp) and developed a mass transfer correlation which is based on impeller characteristics and mixing time but which does not include rheological properties. Loucaides and McManamey [38] also studied oxygen transfer to simulated (with paper pulp) mycelial culture media and while they did characterize the non-Newtonian suspension rheology (with a capillary viscometer) they did not incorporate the rheological properties into their correlation.

In order to develop reliable oxygen transfer correlations it will be necessary for us to obtain more information regarding the quantitative relationship between oxygen transfer rates, mixing conditions and rheological properties. Also required is more information about (1) the effects on mass transfer of morphology and cell clumping (e.g., pellet-like growth), (2) the effects of agitation in breaking up aggregates of cells to increase mass transfer surface area and to decrease "solid" phase diffusion resistance and (3) cell degradation in high shear stress flow fields.

5.2.2. Culture Fluids Containing Microbial Polysaccharides

To the best of the author's knowledge there have been no correlations developed for oxygen transfer to culture fluids containing polysaccharides. However, a basis for such work may have been provided by Perez and Sandall [125] who studied rates of carbon dioxide transfer to solutions of carbopol (0–1.0%) in standard agitated vessels. (Solutions of carbopol (carboxy polymethylene) obey the power law equation and have flow behavior indicies varying from 0.916 for a 0.25% solution to 0.594 for a 1.0% solution.) Yagi and Yoshida [126] extended this work by considering carbon dioxide transfer to other non-Newtonian solutions (sodium polyacrylate and carboxymethyl cellulose). They found that carbon dioxide transfer rates to various solutions having a broad range of rheological properties (including Newtonian) could be correlated by the expression

$$\frac{K_L a_G d_i^2}{D_L} = 0.06 \left(\frac{d_i^2 N \rho}{\eta_a}\right)^{1.5} \left(\frac{d_i N^2}{g}\right)^{0.19} \left(\frac{\eta_a}{\rho D_L}\right)^{0.5}$$

$$\times \left(\frac{\eta_a V_s}{\sigma}\right)^{0.6} \left(\frac{N d_i}{V_s}\right)^{0.32} \times [1 + 2.0 \, (\lambda N)^{0.5}]^{-0.67}, \tag{63}$$

where

$K_L a_G$ = volumetric liquid phase mass transfer coefficient (s^{-1})
η_a = apparent viscosity $(g \, cm^{-1} \, s^{-1})$

D_L =   liquid phase diffusivity (cm^2/s)
σ =   surface tension ($g\,s^{-2}$)
V_s =   superficial gas velocity ($cm-s^{-1}$)
ρ =   liquid density (g/cm^{-3}).

The apparent viscosity η_a was determined by the method of Metzner and Otto [24]. The term (λN) is called the Deborah number [127] and is a measure of viscoelasticity. Some measurements of oxygen transfer rates to xanthan-containing culture fluids have been performed recently in the author's laboratory [128]. Operating conditions were the same as those described for the measurement of mixing times (see Sect. 4.2.3.). The results, some of which are given in Fig. 31, illustrate the profound effect of xanthan-solution rheological properties even at relatively low concentrations. Furthermore, although the data are by no means extensive enough to permit development of a general correlation, they do seem to demonstrate that there are qualitative as well as quantitative differences between the gas absorption characteristics of mycelial culture fluids and those containing a soluble polymer. These differences are probably a result of the different phenomena which cause each type of fluid to exhibit its own non-Newtonian characteristics.

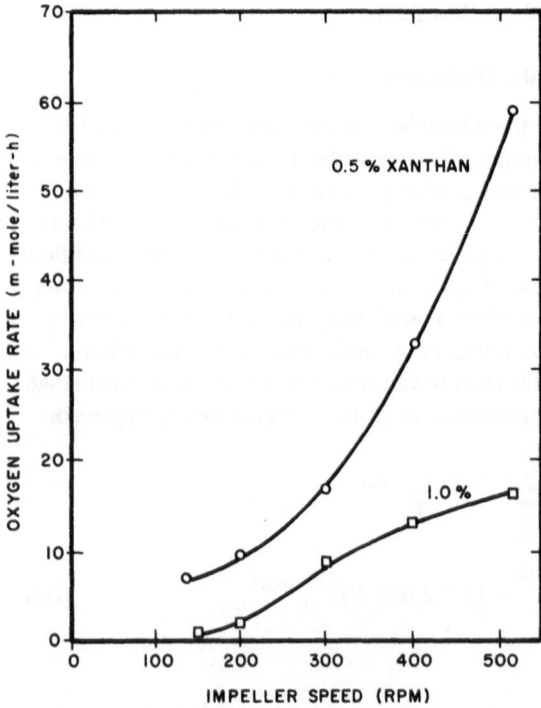

Fig. 31. Oxygen transfer to culture fluids containing xanthan

6. Heat Transfer

Heat Transfer correlations are not available for culture fluids of any kind in stirred tank reactors (aerated or non-aerated). Indeed, the subject is not even discussed in most reviews or biochemical engineering texts. In most cases the available design procedures (particularly for non-Newtonian broths) are inadequate and lead to "conservative over-design" or "retrofitting", both of which are expensive and inefficient.

In the past, heat transfer costs were frequently a minor consideration and hence design inadequacies were tolerated with relatively minor pain. However, increasing heat transfer costs and the anticipated uses of bioprocesses to produce low cost large volume chemicals and chemical intermediates along with integration of bioprocesses into chemical processes will require the use of more accurate heat transfer design methods typical of those employed in the more highly developed chemical process industries. The interested reader is referred to various reviews for discussions of available correlations and techniques [129–131].

7. Overall Effects of Rheological Properties on Microbial Reactions

7.1. Batch Culture

7.1.1. Yield and Productivity Losses

In designing, operating, or analyzing process data from any reactor used to process viscous non-Newtonian fluids, one must always deal with the problems and uncertainties imposed by heterogeneous environments (poor mixing) and depressed rates of heat and mass transfer which, if permitted to persist, lead invariably to (1) lower yields, (2) inferior process control, (3) unreliable process analysis and (4) complex, costly, and frequently unreliable design and scale-up methods. However, despite the fact that the biotechnologists have long recognized the importance of these factors in bioprocesses, they appear to have given them very little quantitative consideration.

To focus attention on the problems alluded to above, we consider the hypothetical fermentation illustrated in Fig. 32. The organism is assumed to be a facultative anaerobe which produces the desired product, A, when the dissolved oxygen concentration (DO) exceeds 15% of saturation. When $DO < 15\%$, the organism produces B which (1) has physical properties similar to A, (2) is inhibitory to the production of A even under aerobic conditions and (3) starts to become lethal at concentrations in excess of 2%. It is further assumed that (1) once the organism switches its metabolism to produce B it can not, for all practical purposes, switch back, (2) mutates spontaneously at low DO to a form which does not produce A or B but which consumes substrates more rapidly than does the parent strain (under conditions of low DO) and (3) the parent strain can not tolerate pH extremes for even short periods of time. Region V_1 is well mixed and is supplied adequately with oxygen while region V_2 is essentially stagnant and has a dis-

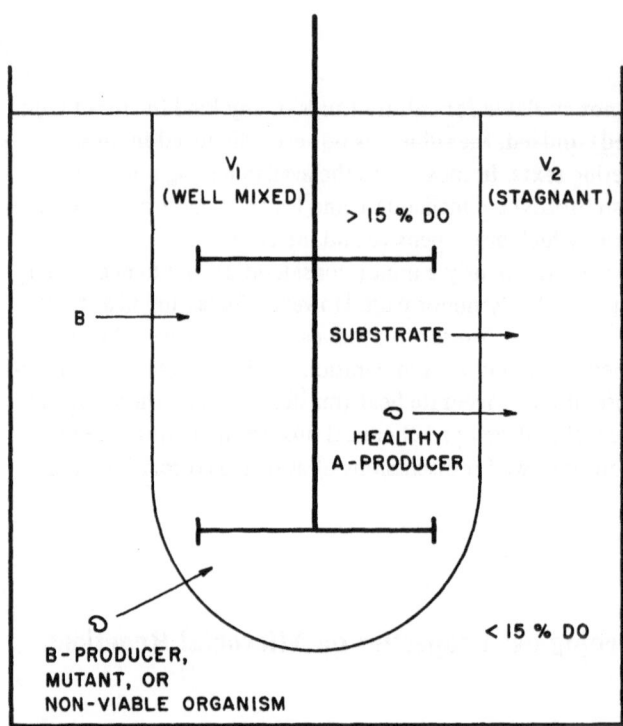

Fig. 32. Model of poorly mixed reactor

solved oxygen concentration of less than 15% of saturation. Such mixing conditions have been discussed for fungal cultures by Phillips and Johnson [94], Maxon [95], and Steel and Maxon [96, 99] and for polysaccharide cultures by LeDuy and Zajic [132] and Charles *et al.* [20].

We observe first that yield and productivity will be lower than they could be (if mixing were adequate) for the simple reason that the effective working volume has been reduced. However, further loss can occur due to the production of B in region V_2. First, since substrate is consumed, a concentration gradient might be established which would cause diffusion of substrate from V_1 to V_2 (obviously this will depend on relative rates of substrate consumption in V_1 and V_2). Second, B will diffuse from V_2 to V_1 and decrease the metabolic rate in V_1. Finally, the presence of B in the final culture fluid may increase the cost of recovering and purifying A and decrease the effective yield and productivity. Movement of organisms between V_1 and V_2 also will lead to losses in productivity and yield. Organisms moving from V_2 to V_1 will not contribute to production and may consume valuable substrate. (It is also possible that they will continue to produce B for a while.) Organisms moving from V_1 to V_2 may (1) be killed if they can not adapt quickly enough to a relatively high concentration of B, (2) switch their metabolism to produce B or (3) mutate. In any event, if they return to V_1 they will no longer be capable of producing A. In this regard it should be noted that the author [133] has observed that during xanthan production there are higher concentrations of non-producing mutants (small colony type [134]) in the stagnant regions of the fermentor than in the well mixed section.

Poor mixing quality and low rates of mass transfer also contribute to inferior process control which results in further losses. Control (e.g., acid or base addition) is exercised on the basis of the environment "seen" by a sensor which in this case can depend strongly on the sensor location. For example, both LeDuy and Zajic [132] and the author [133] have found that during polysaccharide production pH measurements in stagnant and well-mixed regions are not the same. Also, Phillips and Johnson [94], Maxon [95], and Steel and Maxon [96, 97] reported spatial variations of DO during fungal fermentations. Therefore, control action may be taken which is appropriate only for the sensor location. Furthermore, the response of the system to the control action will not be uniform; response will be more rapid in V_1 than in V_2. Finally, sensor response *per se* will also depend on sensor location; diffusional resistance will be greater in V_2 than in V_1 and therefore sensor response will be slower in V_2. One need only imagine the situation of a pH probe located in V_2 and reading a pH a few tenths of a unit different than the pH in V_1 to realize the potential for the creation of environmental extremes which might not be tolerated by the productive population in V_1.

7.1.2. Data Interpretation

Our example has stressed only losses in yield and productivity. However, it is equally important to consider the operating data and the uses to which this data will be put. Clearly, data obtained under the conditions of the example represent metabolic processes which vary both spatially and temporally and are at best difficult to interpret unambigously even if samples are taken at various locations as suggested by LeDuy and Zajic [132]. Indeed, such data may be responsible for many misinterpretations, assorted mysteries and various "anomalies" and may account for many disagreements in the literature. When one couples this notion with the fact that the usual scale up techniques do not involve explicit consideration of mixing characteristics one is not surprised at the inadequacy of reaction design techniques for non-Newtonian processes.

7.1.3. Improvement of Batch Culture Operation

If biotechnology is to compete in the bulk chemicals arena as has been suggested by many prominent biotechnologists [135], it will be necessary to cope seriously with the problems discussed above. The philosophy one adopts to do this will certainly depend on the extent to which he is committed to standard reaction design and operation. Those wishing to retain as much of the "proven" technology as possible might be led to (1) develop monitoring systems capable of providing accurate continuous descriptions of spatial and temporal variations which could be used as bases for scale-up and control strategies and (2) to design better mixing hardware and/or to institute better mixing practices.

7.1.3.1. Improved Monitoring/Control Systems

Control and monitoring problems discussed previously should lead one to recognize
the difficulties associated with (1) above. To monitor adequately important process
variables, probes and sample ports would be required at several positions. But how
many are needed? Where should they be located? How should one weight the responses
for control purposes? For purposes of analysis and/or scale-up? How do these factors
vary with fermentation time? How do system and probe responses vary as a function of
position and reaction time? Little consideration has been given to such questions and
probably for good reason. While they are certainly challenging their solutions probably
would be of little practical value for even if a monitor/control system could be designed
properly it would be expensive to build and to maintain, cumbersome to use, and
plagued continually by questions of long-term reliability. For more basic work these
objections might not be as important but the problem of scale-up would remain. Finally,
it is likely that at best such systems would provide nothing more than better control
and operating data: They would probably do little to improve decreased yield and productivity caused by poor mixing. Therefore, the development of involved monitor/con-
trol systems does not appear to be a promising approach. Providing a more homogeneous
environment by means of improved impeller design, changes in baffling, and/or
operating at greater impeller speeds seems to be the more effective strategy.

7.1.3.2. Improved Agitation

It was noted previously (Sect. 4.3.) that the turbine impeller is probably not the best
choice for mixing non-Newtonian culture fluids. However, only Steel and Maxon [97]
have discussed other designs in detail and the turbine seems to remain the impeller of
choice for most applications at almost all scales of operation. The imposition of this
constraint is indeed unfortunate but still permits improvement of mixing through the
use of higher impeller speeds, smaller T/d_i ratios, multiple impellers, and baffle modification or removal. For mycelial cultures there have been many reports of such approaches,
but useful correlations have not been developed probably because little was done to
study the effects of rheological properties or to quantify mixing characteristics.
The application of these strategies to polysaccharide cultures has not been studied extensively. Rogovin et al. [136] found that increased impeller speed alone gave greater yields
and productivities of xanthan. They gave no information concerning power consumption. The author [63, 64] was able to improve both yield and productivity of xanthan
while attaining increased final gum concentration in a standard 7-liter fermentor by
replacing the standard 6-blade impellers (2) having T/d_i ratios of 2.5 with three pitched-
blade turbines (6-blade, 45° pitch) having T/d_i ratios of 1.8 and by removing the baffles
(some baffling was still provided by probes and reactor internals). The improved performance was obtained at a cost of approximately 75% increased power consumption.
It is interesting to contrast this last observation with the finding of Steel and Maxon
[97] that in the case of a fungal fermentation, the "multiple-rod" impeller performed
as well as the turbine but required substantially less power.
In considering new impeller designs and mixing techniques one must consider not only
the performance/cost ratio but must also take into account the fact that most viscous

non-Newtonian cultures begin as relatively low viscosity Newtonian fluids. In addition, mixing, mass transfer, and heat transfer requirements change dramatically as rheological properties change not only because of rheological effects *per se* but also because metabolic patterns change. Perhaps it is expecting too much to believe that a single impeller can satisfy all the requirements. If this is the case, one might consider using an impeller whose configuration could be changed automatically or incorporating more than one type of impeller in a single vessel. Certainly, both of these can be done but only with considerable and undesirable mechanical complexity. Alternatives which should be considered seriously but which can not be discussed here include (1) staging of batch systems and (2) use of other types of reactors (e.g., tubular reactors).

7.1.3.3. The Dilution Method

Another approach which has met with some success is the "dilution technique": When the rheological properties of the culture cause an unacceptable decrease in mixing quality the culture is diluted with sterile water or fresh medium (essentially a fed-batch operation) in order to decrease viscosity and non-Newtonian behavior and thereby to improve mixing. Satoh [137] reports that in the Kanamycin process a 10% dilution resulted in a 50% viscosity reduction and increased product yield. Others [109] have suggested more general application but while the method does seem to have merit there is little in the literature to support general applicability. Certainly, careful analyses of the effects of dilution on yield and productivity are required for a range of culture fluids exhibiting a variety of rheological properties. For example. in the case of xanthan a dilution much greater than 10% would be required to effect a viscosity decrease of 50% (particularly at high xanthan concentrations) and even then the non-Newtonian properties (e.g., the power law index) would not change drastically. The large dilution required probably would result in yield and productivity losses which would outweigh gains resulting from better mixing. The results of initial small-scale experiments (1–5 l) performed in the author's laboratory substantiate this contention in the case of xanthan [138].

7.2. Continuous Culture

7.2.1. General Considerations

There have been many reported studies of laboratory and small-scale continuous fermentations involving non-Newtonian mycelial cultures but in none of these were the effects of rheological properties and mixing conditions treated in a quantitative fashion despite the fact that high viscosity and pronounced non-Newtonian behavior appear to cause significant departure from theoretical predictions based on batch data and the assumption of perfect mixing in the continuous reactor. Such deviations should not be surprising because (1) as noted previously for such cultures, batch data itself is often difficult to interpret unambiguously, (2) mixing characteristics often are different in batch and continuous reactors used in reported studies and (3) perfect mixing probably was achieved infrequently in the continuous reactors used. It is also important to note that rheological properties and hence mixing characteristics can vary significantly with dilution rate and therefore the internal consistency of results of any given study should be examined closely. The observed departures from theory have led to the generation of a plethora of "modified" equations which address only reaction kinetics *per se* and ignore completely the effects of adverse rheological conditions and poor mixing. This is probably one of the major reasons for disagreements among various investigators and for the lack of acceptable design correlations for non-Newtonian continuous culture. It is also possible that at least in the cases under consideration spontaneous mutation "takeover", which is one of the major problems blocking the acceptance of commercial continuous culture, is promoted in poorly mixed regions where the effective dilution rate is much lower than the average dilution rate and conditions are possibly more faborable for successful competition by mutants (see Sect. 7.1.1. for comments regarding a similar situation in batch fermentors). At the same time, the effective dilution rate in the well mixed region is greater than the average and this can contribute to unstable operation. Oscillatory behavior observed during a number of continuous processes may be associated with interchanges between stagnant and well-mixed regions or with shifts in the locations of stagnant regions. Oscillations following step changes in operating conditions may also result, in part, from mixing effects.

Certainly, much of the preceeding discussion is speculative but it does stress the importance of mixing and rheological properties in continuous culture and emphasizes the need for further study of these factors. The author feels that techniques developed for the study of mixing effects in chemical reactors should be considered for investigations of continuous cultures.

7.2.2. Residence Time Distribution

Extensive studies of mixing effects in continuous chemical reactors have led to useful techniques for the analyses of experimental results and for the design of commercial reactors. Among those which may prove to be of value in analyzing and designing bioreactors is the *residence time distribution* (RTD) concept [76–78]. The *residence time*, τ, of a particle is simply the actual time it spends in the reactor. The *residence time distribution function, $J(\tau)$*, is the fraction of particles in the effluent that have residence times less than τ. $J(\tau)$ may be obtained directly from the response of the reactor to a step change in inlet concentration as shown in Fig. 33:

$$J(\tau) = \frac{C(\tau)}{C_0} \tag{64}$$

where $C(\tau)$ is the exit concentration and C_0 is the inlet concentration of the tracer. Another function which should prove valuable is the *internal age distribution function, $I(\tau)$* which is defined in such a manner that $I d\tau$ is the fraction of particles (in the vessel) having ages between τ and $\tau + d\tau$. Thus

$$\int_0^\infty I d\tau = 1. \tag{65}$$

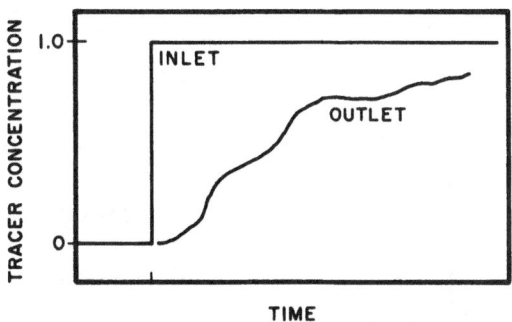

Fig. 33. Schematic response of a continuous stirred tank reactor to a step change in tracer input concentration

Furthermore, it can be shown [123] that

$$I(\tau) = 1 - J(\tau). \tag{66}$$

These and other age distribution functions can be used in conjunction with a variety of mixing models to simulate reactor performance [137, 138]. The application of these techniques to bioreactors used to process non-Newtonian cultures will be difficult and will require simultaneous consideration of mixing, mass transfer, and microbial kinetics with the recognition that the latter two factors may be spatially dependent. However, this should prove to be a challenging and rewarding endeavor. For a further description of mixing effects on the performance of continuous fermentors the reader is referred to the discussion and review in the new textbook by Bailey and Ollis [78].

To the author's knowledge studies of reactor residence time distributions have been limited to processes involving yeasts and bacteria. Most of this work was associated with investigations of multistage tower reactors and is adequately reviewed by Bailey and Ollis [78]. The only other reported studies even marginally related to bioprocesses are those of Hubbard and Calvetti [139] and Johnson and Hubbard [140] who discussed residence time distribution applications in reactions and reported experimental $J(\tau)$ curves for agitated vessels containing solutions of xanthan gum. Their results are quite similar to those obtained by the author. In Fig. 34 a typical RTD for a 0.5% xanthan solution is compared with the RTD for a perfectly mixed vessel (replotted from results of Hubbard and Calvetti [129]—see original paper for details). The departure from ideal behavior is striking and helps to illustrate the need for more work in this area.

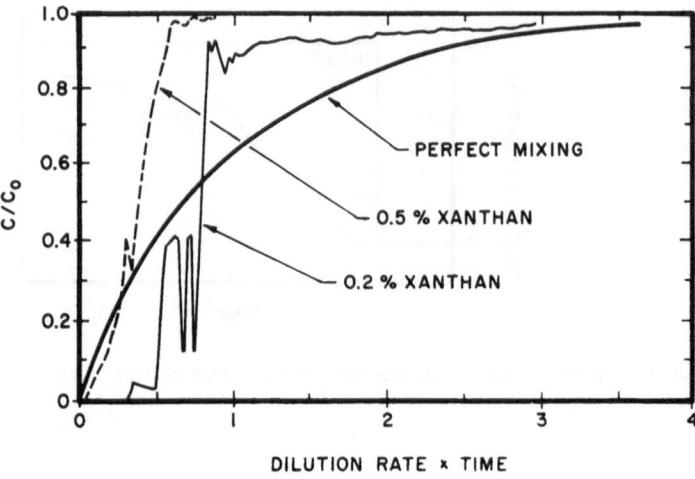

Fig. 34. Response of a continuous stirred tank reactor containing xanthan solutions to a step change in tracer input concentration. (Replotted from results of Hubbard and Calvetti [139])

8. Rheological Properties for Monitoring and Control Applications

The pronounced variations of the rheological properties of a non-Newtonian culture during a fermentation suggest the use of rheological measurements for monitoring/control applications. While the potential value of this approach has been recognized for some time, little has been done to provide the basic information, tools, and techniques necessary for its implementation. Certainly, "rheological histories" have been determined for a number of fermentations but in none of the reported studies was the information gathered or analyzed in a manner which would make it completely suitable for monitoring/control applications.

In order to use rheological measurements for monitoring/control purposes in a given fermentation it will be necessary to identify first the properties which best characterize and, are most sensitive to changes in, the state of the reaction and which can be measured reproducibly (from one batch to the next) and with the least effort. These properties should be correlated with cell concentration and morphology in the case of mycelial cultures and with product concentration and molecular weight (and possibly molecular weight distribution) in the case of polysaccharide fermentations. In pursuing this goal one should certainly consider parameters which describe the shear dependence of viscosity (e.g., K and m of the power law) but should not overlook other properties such as yield stress, thixotropy, and viscoelasticity as these can be much more sensitive indicators than viscosity parameters.

It would be most desirable if the rheological measurements could be made continuously and *in situ*. However, the only report of *in situ* viscosity measurement during a fermentation is that of Wang und Fewkes [48] and, on the basis of comments made in Sect. 2.4.2.5., there is some doubt regarding the suitability of their method for monitoring or control applications (it does not measure rheological properties directly nor under conditions satisfying the definition of viscosity) but it certainly warrants further examinations. As an alternative one might consider experimenting with a device such as the Dynatrol (Sect. 2.4.2.5.) or developing for use *in situ* an instrument which actually measures viscosity in accord with the accepted definition.

9. Concluding Remarks

The rheological characteristics of culture fluids and their effects on bioprocesses have not been studied to the extent or with the rigor warranted by their importance. Rheological properties influence profoundly the various transport phenomena and, in turn, these affect directly our ability to maintain a controlled homogeneous environment which provides us a mean of controlling the response of the organism. In the past, many of the ambiguities, misinterpretations, inadequate design correlations and methods, suboptimal control strategies, and low yields and productivities, resulted, in large part, because the aforementioned interactions were either ignored completely, or were not treated in a quantitative fashion based on both microbiological and physical data obtained during the same experiment. To overcome these problems and to develop the accurate intrinsic kinetic models, the fermentor design correlations and practices, and the operating and control protocols that will be required for a new generation of reactors (not necessarily the classical stirred-tank) suitable for use in the chemical process industries, every effort must be made to study rheological properties, transport phenomena, and microbiological behavior simultaneously in order to elucidate quantitatively their interactions. Certainly, the highly instrumented computer coupled systems now available should provide the means for doing this.

Also, it is worth re-emphasizing that not only has the study of culture broth rheology *per se* been grossly neglected, but methods and instruments (particularly for *in situ* measurement) have not been tested and characterized adequately. Efforts to develop reliable and meaningful methods and instruments for measurement of a variety of rheological properties will be worthwhile not only for the reasons stated above, but also because for most non-Newtonian cultures several rheological properties (in addition to viscosity) should be quite useful for characterizing the state of a culture and hence for monitoring/control applications. Clearly, *in situ* measurements would be most desirable for such applications.

Acknowledgement

Portions of this work were sponsored by the Pennsylvania Science and Engineering Foundation under PSEF Agreement # 273.

Nomenclature

a_g	specific surface area, gas phase	L^{-1}
a_s	specific surface area, cell	L^{-1}
C_A	substrate concentration,	ML^{-3}
C_A^*	substrate concentration, equilibrium	ML^{-3}
C_{AS}	substrate concentration, cell surface	ML^{-3}
D	diameter	L
D_L	diffusivity	$L^2\theta^{-1}$
d_i	impeller diameter	L
g	gravitational constant	$L\theta^{-2}$
h	liquid level	L
$I(\tau)$	internal age distribution function	–
$J(\tau)$	residence time distribution function	–
K	consistency index (power law model; pseudopoise)$_\tau$	$ML^{-1}\theta^{m-2}$
K_L	overall mass transfer coefficient (liquid phase)	$L\theta^{-1}$
K_G	overall mass transfer coefficient (gas phase)	$L\theta^{-1}$
k_1	first order rate constant	θ^{-1}
k_L	liquid film mass transfer coefficient	$L\theta^{-1}$
L	length	L
M	torque	$ML^2\theta^{-2}$
$\dot{M}_{A_g}$	substrate flow rate, gas film (per unit volume)	$ML^{-3}\theta^{-1}$
$\dot{M}_{A_l}$	substrate flow rate, liquid film (per unit volume)	$ML^{-3}\theta^{-1}$
Mf	momentum factor	–
m	power law index	–
N	rotational speed	θ^{-1}
N_m	mixing number	–
N_p	power number	–
N_Q	volumetric discharge number	–
N_{Re}	Reynolds number	–
P	power	$L^2M\theta^{-3}$
ΔP	pressure difference	$M\theta^{-2}$
R, r	radius	L
R_i	inner radius	L
R_o	outer radius	L
r_A	specific overall reaction rate	$ML^3\theta^{-1}$
r_A	specific chemical reaction rate	$ML^3\theta^{-1}$
T	reactor diameter	L
u	fluid velocity	$L\theta^{-1}$
u_m	maximum fluid velocity	L^{-1}
$\bar{u}$	average fluid velocity	$L\theta^{-1}$
V	liquid volume	L^3
V_s	superficial gas velocity	$L\theta^{-1}$
V_z	axial velocity	$L\theta^{-1}$
W	impeller blade width	L
X	cell mass concentration	ML^{-3}
y	liquid height	L
α	cone angle	–
$\dot{\gamma}$	shear rate	θ^{-1}
$\dot{\gamma}_w$	shear rate at wall	θ^{-1}
η	Newtonian viscosity	$ML^{-1}\theta^{-1}$
η_a	apparent viscosity	$ML^{-1}\theta^{-1}$
η_p	plastic viscosity (rigidity)	$ML^{-1}\theta^{-1}$
θ	time	θ^{-1}

θ_m	mixing time	θ^{-1}
λN	Deborah number	–
μ_{ap}	field apparent viscosity	$ML^{-1}\theta^{-1}$
ν	kinematic viscosity	$L^2\theta^{-3}$
ρ	density	ML^{-3}
σ	surface tension	$M\theta^{-2}$
τ	shear stress	$ML^{-1}\theta^{-2}$
τ_y	yield stress (Bingham plastic)	$ML^{-1}\theta^{-2}$
τ_w	shear stress at wall	$ML^{-1}\theta^{-2}$
Ω, ω	angular speed	θ^{-1}

References

1. Van Wazer, J. R., Lyons, J. W., Lim, K. Y., Colwell, R. E.: Viscosity and Flow Measurement, p. 23. New York: Interscience 1963.
2. Bird, R. B., Stewart, W. E., Lightfoot, E. N.: Transport Phenomena, Chap. 3. New York: Wiley 1960.
3. Fredrickson, A. G.: Principles and Applications of Rheology. Englewood Cliffs, N. J.: Prentice-Hall, 1964.
4. Wilkinson, W. L.: Non-Newtonian Fluids, pp. 130–134. New York: Pergamon 1960.
5. Skelland, A. H. P.: Non-Newtonian Flow and Heat Transfer, p. 40. New York: Wiley (1967).
6. *Ibid.* p. 39–47.
7. Middleman, S.: The Flow of High Polymers, Chap. 2. New York: Interscience 1968.
8. Krieger, I. M., Maron, S. H.: J. Appl. Phys. **23**, 147 (1952).
9. *Ibid.* **24**, 134 (1953).
10. *Ibid.* **25**, 72 (1954).
11. Metzner, A. B., Taylor, J. S.: AIChE J. **6**, 109 (1960).
12. Wilkinson, W. L.: Non-Newtonian Fluids, p. 5. New York: Pergamon 1960.
13. Bird, R. B., Stewart, W. E., Lightfoot, E. N.: Transport Phenomena, New York: Wiley 1960.
14. Wilkinson, W. L.: Non-Newtonian Fluids. New York: Pergamon 1960.
15. Fredrickson, A. G.: Principles and Applications of Rheology. Englewood Cliffs, N. J.: Prentice-Hall 1964.
16. Skelland, A. H. P.: Non-Newtonian Flow and Heat transfer. New York: Wiley 1967.
17. Wilkinson, W. L.: Non-Newtonian Fluids, pp. 3, 53–55. New York: Pergamon 1960.
18. Roels, J. A., Van Den Berg, J., Voncken, R. M.: Biotechn. Bioeng. **16**, 181 (1974).
19. Wilkinson, W. L.: Non-Newtonian Fluids, p. 6–9. New York: Pergamon 1960.
20. Charles, M., Toth, G. M.: paper presented at the 168th Annual Meeting of the American Chemical Society, Atlantic City, N. J., Sept. 8–13, 1974.
21. Charles, M.: Unpublished data.
22. Ulbrecht, J.: The Chem. Eng., June 1974, 347.
23. Metzner, A. B., Otto, R. E.: AIChE J. **3**, 3 (1957).
24. Foresti, R., Liu, T.: IEC **51**, 860 (1950).
25. Calderbank, P. H., Moo-Young, M. B.: Trans. Inst. Chem. Eng. **37**, 22 (1959).
26. *Ibid.* **39**, 22 (1961).
27. Skelland, A. N. P.: Non-Newtonian Flow and Heat Transfer, Chap. 5. New York: Wiley 1967.
28. Van Wazer, J., Lyons, J. W., Lim, K. Y., Colwell, R. E.: Viscosity and Flow Measurement, New York: Interscience 1963.
29. Oka, S.: In: Rheology, Theory and Applications, 3. Eirich, F. R. (Ed.), New York: Academic Press 1960.
30. Van Wazer, J. R., Lyons, J. W., Lim, K. Y., Cowell, R. E.: Viscosity and Flow Measurement, pp. 199–214. New York: Interscience 1963.

31. Skelland, A. P. H.: Non-Newtonian Flow and Heat Transfer, p. 32–36. New York: Wiley 1967.
32. Middleman, S.: The Flow of High Polymers, p. 15–18. New York: Interscience 1968.
33. Rabinowitsch, B.: Z. Physik. Chem. A **145**, 1 (1929).
34. Mooney, M.: J. Rheol. **2**, 210 (1931).
35. Van Wazer, J. R., Lyons, J. W., Lim, K. V., Cowell, R. E.: Viscosity and Flow Measurement, Chap. 4. New York: Interscience 1963.
36. Skelland, A. H. P.: Non-Newtonian Flow and Heat Transfer, Chap. 3. New York: Wiley 1967.
37. Goldsmith, H. L., Mason, S. G.: In: Rheology, Theory and Application, Vol. 4, p. 215–220, Eirich, F. R. (Ed.) New York: Academic Press 1967
38. Loucaides, R., McManamey, W.J.: Chem. Eng. Sci. **28**, 2165 (1973).
39. Van Wazer, J. R., Lyons, J. W., Lim, K. Y., Cowell, R. E.: Viscosity and Flow Measurement, p. 139–150. New York: Interscience 1963.
40. Skelland, A. H. P.: Non-Newtonian Flow and Heat Transfer, p. 47. New York: Wiley 1967.
41. Middleman, S.: The Flow of High Polymers, p. 25–28. New York: Wiley 1968.
42. Bongenaar, J. J. T. M., Kossen, N. W. F., Metz, B., Meijboom, F. W.: Biotech. Bioeng. **15**, 201 (1973).
43. Charles, M.: Unpublished data.
44. Charles, M.: Unpublished data.
45. Van Wazer, J. R., Lyons, J. W., Lim, K. Y., Cowell, R. E.: Viscosity and Flow Measurement, p. 150–156. New York: Interscience 1963.
46. Humphrey, A. E.: Paper presented at the Labex Symposium on Computer Control of Fermentation Processes, Labex International, Earls Court, London, 1971.
47. Nyiri, L. K.: *ibid.*
48. Fewkes, C. J., Wang, D. I. C.: Paper presented at First Chemical Congress of the North American Continent, Mexico City, Mexico, Nov. 30–Dec. 5, 1975.
49. Morris, G. G., Greenshields, R. N., Smith, E. L.: Biotechn. Bioeng. Symp. 4, 535 (1963).
50. Solomons, G. L., Weston, G. O.: Biotech. Bioeng. **1**, 1 (1961).
51. Deindoerfer, F. H., Gaden, E. L.: Appl. Micro. 3, 253 (1955).
52. Deindoerfer, F. H., West, J. M.: Biotech. Bioeng. **2**, 165 (1960).
53. Richards, J. W.: Prog. Ind. Micro. 3, 141 (1961).
54. Tuffile, C. M., Pinho, F.: Biotech. Bioeng. **12**, 849 (1970).
55. Taguchi, H., Miyamoto, S.: Biotech. Bioeng. **8**, 43 (1966).
56. Charles, M.: Unpublished data.
57. Charles, M.: Unpublished data.
58. Jeanes, A., Pittsley, J. E.: J. Appl. Poly. Sci. **17**, 1621 (1973).
59. Kelco Co.: Xanthan Gum (1975).
60. Patton, T. C.: J. Paint Tech. **38**, 656 (1966).
61. Patton, T. C.: Cereal Sci. Today, **14**, 178 (1969).
62. LeDuy, A., Marsan, A. A., Coupal, B.: Biotech. Bioeng. **16**, 61 (1974).
63. Charles, M., Radjai, M. K.: Paper presented at the First International Congress on Engineering and Food, Aug. 9–14, 1976, Boston, Mass.
64. Charles, M., Radjai, M. K.: Paper presented at the 172nd ACS National Meeting, Symposium on Extracellular Microbial Polysaccharides of Industrial Importance, San Francisco, CA, Aug. 31, 1976.
65. Rogovin, S. P., Anderson, R. F., Cadmus, M. C.: Biotechn. Bioeng. **1**, 51 (1961).
66. Moraine, R. A., Rogovin, S. P.: Biotechn. Bioeng. **13**, 381 (1971).
67. Knittig, E., Zajic, J. E.: Biotech. Bioeng. **14**, 379 (1972).
68. Kosaric, N., Yu, J. E., Zajic, J. E.: Biotech, Bioeng. **15**, 729 (1973).
69. Rogovin, S. P., Sohno, V.E., Griffin, E. L.: IEC **53**, 37 (1961).
70. Burton, K. A., Cadmus, M. C., Lagoda, A. A., Sanford, P. A., Watson, P. R.: Biotech. Bioeng. **18**, 1669 (1976).
71. Deindoerfer, F. H., West, J. M.: Adv. Appl. Micro. **2**, 265 (1960).
72. Eirich, F.: Kolloid Z. **74**, 276 (1936).

73. Shimmons, B. W., Svrcek, W. Y., Zajic, J. E.: Biotech. Bioeng. **18**, 1793 (1976).
74. Modeer, B.: Proc. Biochem., Sept. 1974, 23.
75. Goldsmith, H. L., Mason, S. G.: In: Rheology, Theory and Applications, Vol. 4. New York: Academic Press (1967).
76. Levenspiel, O.: Chemical Reaction Engineering, Chaps. 9, 10. New York: Wiley 1962.
77. Smith, J. M.: Chemical Engineering Kinetics, Chap. 6. New York: McGraw-Hill 1970.
78. Bailey, J. E., Ollis, D. F., Biochemical Engineering Fundamentals, Chap. 9. New York: McGraw-Hill (in press)
79. Nagata, S.: Mixing: Principles and Applications, p. 194. New York: Wiley 1975.
80. Novak, V., Riegler, F.: Che. Eng. J. **9**, 63 (1975).
81. Moo-Young, M., Tichar, K., Dullien, F. A.: AIChE J. **18**, 178 (1972).
82. Nagata, S., Nishikawa, M., Katsube, T., Takaish, K.: Int. Inl. Che. Eng. **12**, 715 (1972).
83. Hoogendorn, C. J., Den Hartog, A. P.: CES **22**, 1689 (1967).
84. Norwood, K. W., Metzner, A. B.: AIChE J. **6**, 432 (1960).
85. Fox, E. A., Gex, V. E.: AIChE J. **2**, 539 (1956).
86. Godleski, E. S., Smith, J. C.: AIChE J. **8**, 617 (1962).
87. Khang, S. J., Levenspiel, O.: Chem. Eng., Oct. 11, 1976, 141.
88. Nagata, S.: Mixing: Principles and Applications, p. 204. New York: Wiley 1975.
89. *Ibid.* p. 206.
90. Hicks, R. W., Morton, J. R., Fenic, J. G.: Chem. Eng., Apr. 26, 1976.
91. Nagata, S.: Mixing: Principles and Applications, p. 200. New York: Wiley
92. Wang, D. I. C., Humphrey, A. H.: In: Progress in Industrial Microbiology, Vol. 8. Hockenhull, J. J. D. (Ed.) Cleveland: CRC Press 1968.
93. Hyman, D.: In: Advances in Chemical Engineering, Vol. 3, p. 120. Drew, T. B., Hoopes, J. W., Vermeulen, T. (Ed.) New York: Academic Press (1962).
94. Phillips, D. H., Johnson, M. J.: Biotech. Bioeng. **3**, 277 (1961).
95. Maxon, W. D.: Biotech. Bioeng. **1**, 311 (1959).
96. Steel, R., Maxon, W. D.: Biotech. Bioeng. **8**, 97 (1966).
97. Steel, R., Maxon, W. D.: Biotech. Bioeng. **8**, 109 (1966).
98. Blakebrough, N., Sambamurthy, K.: Biotech. Bioeng. **8**, 25 (1966).
99. Wang, D. I. C., Fewkes, R. C. J.: Paper presented at the Thirty Second Meeting of the Society of Industrial Microbiology, Jekyll Island, Georgia, Aug. 16–20, 1976.
100. Charles, M.: Unpublished data.
101. Leamy, G. H.: Chem. Eng., Oct. 15, 115 (1973).
102. Charles, M.: Unpublished data.
103. Chavan, V., V., Arumugam, M., Ulbrecht, J.: AIChE J. **21**, 613 (1975).
104. Nagata, S.: Mixing: Principles and Applications, Chap. 1. New York: Wiley 1975.
105. Blanch, H. W., Bhavaraju, S. M.: Biotech. Bioeng. **18**, 745 (1976).
106. Wilkinson, W. L.: Non-Newtonian Fluids, Chap. 5. New York: Pergamon 1960.
107. Metzner, A. B.: In: Advances in Chemical Engineering, 1. Drew, T. B., Hoopes, J. W., Vermeulen, T. (Ed.) New York: Academic Press 1956.
108. Metzner, A. B., Feehs, R. H., Ramos, H. L., Otto, R. E., Tuthill, J. D.: AIChE J. **7**, 3 (1961).
109. Taguchi, H.: In: Advances in Biochemical Engineering, Vol. 1. Ghose, T. K., Fiechter, A. (Ed.) Berlin, Heidelberg, New York: Springer 1971.
110. Sherwood, T. K., Pigford, R. L., Wilke, C. R.: Mass Transfer. New York: McGraw-Hill 1975.
111. Treyball, R. E.: Mass Transfer Operations. New York: McGraw-Hill 1968.
112. Geankoplis, C. J.: Mass Transport Phenomena. New York: Holt, Rinehart and Winston 1972.
113. Astarita, G.: Mass Transfer With Chemical Reactor. New York: Elsevier 1967.
114. Johnson, D. L., Saito, H., Polejes, J. D., Hougen, O. A.: AIChE J. **3**, 411 (1957).
115. Calderbank, P. H., Jones, S. J. R.: Trans. Inst. Che. Eng. **39**, 363 (1961).
116. Miller, D. N.: IEC **56** (10), 18 (1964).
117. Miller, D. N.: IEC Proc. Des. Dev. **10**, 365 (1971).
118. Miura, Y.: In: Advances in Biochemical Engineering, Vol. 4, Ghose, T. K., Fiechter, A., Blakebrough, N. (Eds.) Berlin, Heidelberg, New York: Springer 1976

119. Atkinson, B., Daoud, I. S.: *ibid.*
120. Atkinson, B., Fowler, H. W.: *ibid.*, Vol. 4 (1976).
121. Smith, J. M.: Chemical Engineering Kinetics, Chaps. 8–12. New York: McGraw-Hill 1970.
122. Carberry, J. J.: Chemical and Catalytic Reaction Engineering, Chaps. 8–10, New York: McGraw-Hill 1976.
123. Charles, M.: Unpublished data.
124. Chain, E. B., Gaulandi, G., Morisi, G.: Biotech. Bioeng. 8, 595 (1966).
125. Perez, J. F., Sandall, O. C.: AIChE J. 20, 770 (1974).
126. Yagi, H., Fumitake, Y.: IEC Proc. Des. Dev. 14, 488 (1975).
127. White, J. L., Tokita, N.: J. Appl. Poly. Sci. 11, 321 (1967).
128. M. Charles: Unpublished data.
129. Edwards, M. F., Wilkenson, W. L.: The Chem. Engr., Sept. 1972, 328.
130. Mitsuishi, N., Miyairi, Y.: Inl. Che. Eng. Jap. 6, 415 (1973).
131. Nagata, S.: Mixing: Principles and Applications, Chap. 2, New York: Wiley 1975.
132. LeDuy, A., Zajic, J. E.: Biotech. Bioeng. 15, 579 (1973).
133. Charles, M.: unpublished data.
134. Cadmus, M. C., Burton, K. A., Herman, A. I., Rogovin, S. P.: Bact. Proc. 1971, 447.
135. Powell, A. J., Bu'Lock, J. D. (Eds.): Octagon Papers 2. Manchester, England: University of Manchester 1975.
136. Moraine, R. A., Rogovin, S. P.: Biotech. Bioeng. 15, 225 (1973).
137. Satch, K.: J. Ferm. Techn. 39, 517 (1961).
138. Charles, M.: Unpublished data.
139. Hubbard, D. W., Calvetti, F.: AIChE J. 18, 663 (1972).
140. Johnson, D. N., Hubbard, D. W.: Biotech. Bioeng. 16, 1283 (1974).

Application of Tower Bioreactors in Cell Mass Production

K. Schügerl, J. Lücke
Institut für Technische Chemie der TU Hannover, D-3000 Hannover

J. Lehmann, F. Wagner
Gesellschaft für Biotechnologische Forschung mbH, D-3301 Stöckheim

Contents

Summary

This article considers the applicability of tower bioreactors without mechanical agitation (bubble columns) for cell mass production on alcohol and glucose substrates. The growth of *Candida boidinii* was investigated in one-stage tower reactors for both batch and "extended culture" operations.

Since in the early stage of cell cultivation growth is controlled already solely by the oxygen transfer rate, various aerators and substrates were compared.

Thus three types of aerator and three substrates were investigated under co- and countercurrent-flow conditions. The effects of antifoam agents were studied in the absence of a mechanical foam separator, while either a foam separator or destroyer was included when antifoam agents were not used. To aid the assessment of the tower bioreactors the following properties were investigated: cell productivity, oxygen transfer rate, volumetric mass transfer coefficient, bubble size, specific gas/liquid interfacial area, and energy requirement. The aerator type as well as the substrate type strongly

influenced these properties. To show that not only synthetic culture media can be applied in tower
bioreactors complex media were also investigated.

The results prove the applicability of tower bioreactors to cell mass production. These bioreactors
are especially attractive because of their high oxygen transfer rate and low energy requirement.

1. Introduction

To improve the economy of cultivation processes not only do the industrial microbial
strains need to be improved, but also the construction of the bioreactors and their
operation be improved as well. "Tower reactors" have become increasingly interesting
in bio-technology because of their high oxygen transfer rates at low energy requirement.

The application of this reactor type is hindered by the lack of necessary data for optimal
construction and operation. This paper deals with tower bioreactors, especially with
their application to SCP productions.

The previous article [1] considered the same reactor type, but with SCP model media.
These two reports together give a review on tower fermentors: article [1] without cells
and the present report under microbial growth conditions.

To show that tower reactors other than those with "synthetic media" yield satisfactory
results, some results have been obtained with "complex media".

The results presented were obtained by two research groups, which are carrying out
cooperative research on tower reactors financed by the German Federal Government
as part of its "Bio-technology program".

2. Characterization of Tower Bioreactors

In the foregoing article [1] the fluid dynamics and transport processes in bubble
columns were investigated with model growth media consisting of salts and alcohols as
substrates; for comparison demineralized water and 10% sodium sulphate solution were
used.

It was shown that the most important properties of bubble columns, namely the mean
relative gas hold-up, the specific interfacial area, the volumetric mass transfer coefficient
and/or oxygen transfer rate (OTR) are controlled by the superficial gas velocity and by
the bubble size. The latter is only a function of the dynamical equilibrium bubble diameter which in turn is controlled by the energy dissipation rate and the surface tension in systems with strong coalescence, but the bubble size is a complex function
of the aerator type, the composition of the liquid and the superficial gas velocity in
systems with hindered coalescence. The most important differences between the cultivation model media treated in the foregoing article and the cultivation media with
growing cells considered in this article are
- the presence of cells, i.e., solid particles,
- the presence of surfactants produced by the cells to maintain their optimal environment,

- the presence of cell constituents due to the death of the cells and leakage through cell membranes,
- the presence of antifoam agents,
- the presence of CO_2 and its desorption.

The cells as well as the surfactants, cell constituents (e.g., proteins), and antifoam agents influence the gas/liquid interface, i.e., the liquid surface properties. Furthermore the influence of interfacial phenomena on the behaviour of bubble columns containing cultivation media and on foam formation has to be considered.

2.1. Influence of the Composition of the Cultivation Medium on the Behaviour of the Gas-Liquid Interface

a) *Inorganic salts* applied as solute increase the surface tension. These salts make the whole solution, interior and surface, very much more cohesive or strongly bound together [2]. This effect probably arises from the attraction of solvated ions of opposite charge augmenting other attraction forces so that the total attraction between solvated ions is greater than that between water molecules. In consequence, an increase in the surface tension is obtained when the concentration of solvated ions is large. The value of the ratio

$$\frac{\text{surface tension}}{\log{(\text{concentration})}}$$

is positive thereby indicating that the excess of water molecules and not the solute ions adsorbs at the surface. A freshly cleaved surface of a solution of one of these electrolytes will, in general, possess a surface tension for any given concentration greater than of the surface equilibrium. This is because the concentration of solvated ions in the fresh surface region is at a maximum value and as the surface ages and approaches equilibrium, solvated ions leave the surface to give way to adsorption of water molecules. During this adsorption process the surface tension is diminished from its maximum value to the equilibrium value, which is still greater than that of pure water.

In the presence of salts an electrical double layer forms at the interface [4] which suppresses the coalescence of bubbles [31].

b) *Alcohols* possess a surface tension much lower than that of water. The surface tension of an alcohol-water solution is between these surface tensions of pure water and alcohol and depends on concentrations.

The surface tension of freshly-formed surface is high but diminished quickly and reaches an equilibrium value, because the component with lower cohesion, the alcohol, is squeezed to the outside surface [2]. The lowering of the surface tension of water by ethanol is greater than the lowering by methanol. This is because ethanol is less attracted by water than methanol and shows a greater tendency to orient at a liquid/air surface. The molecule orients itself with the hydrocarbon groups away from the water. At equilibrium the surface region will contain an excess of oriented alcohol molecules. The presence of an excess of molecules at the surface but in random orientation diminishes the surface tension from that of freshly formed surfaces. Orientation of the adsorbed molecules further reduces the surface tension to its equilibrium value. The presence of the dipole oriented molecules at the surface suppresses the coalescence rate of bubbles [32–34]. With increasing length of the hydrocarbon "tail" the degree of enrichment and orientation of the molecules increases, but the solubility decreases. Thus the properties of the interface gradually change to those of "insoluble films" characteristic of surface covered by slightly soluble surfactants.

c) The distinguishing feature of *surface active agents* is that they lower the surface tension of the pure liquid by adsorbing strongly at relatively low bulk concentrations [3]. The surface tension of the freshly-formed surface will be that of pure water, because there is insufficient solute concentration in the bulk to affect the cohesive forces. A freshly-formed surface will be so sparsely populated with surface active agents that it will behave as pure water. With increasing time the surface tension gradually decreases to its equilibrium value.

The presence or the surfactant film lowers the interfacial and/or surface tension which assumes the value σ instead of value σ_0 in the absence of the interfacial phase. By definition the difference $\pi = \sigma_0 - \sigma$ is the surface pressure of the interfacial film. Experimentally it can be shown that there is no slip between the interfacial films and their subphase. Therefore the expansion or flow of surface films generates a significant transport of a portion of the subphase. Surface and interfacial films are practically without exception very compressible. By preparing a film within a frame and varying the area we may conveniently study the compression or expansion of the film and in particular construct its isotherm:

$$\pi = f(A'),$$

where A' is the surface area covered by the film.

In this type of experiment shear strains are in general negligible with respect to dilatational-extensive strains. For insoluble films it is easy to define a compression modulus κ by Eq. (1a):

$$\kappa = -A'\left(\frac{\partial \pi}{\partial A}\right)_T, \tag{1a}$$

which is equal to the reciprocal compressibility.

The two-dimensional compressibility C_s is defined by Eq. (1b):

$$C_s = -\frac{1}{A'}\left(\frac{\partial A'}{\partial \pi}\right)_T. \tag{1b}$$

The modulus $\kappa = C_s^{-1}$ is sometimes called surface elasticity. At high surface pressure the solubility of "insoluble" films can no longer be neglected.

As a result of the film extension the concentration of the surfactant in the surface is decreased. It is replaced by transport into the surface film from the subphase and the surface tension tends to return to its equilibrium value.

That is, by expanding the surface film more surfactant is adsorbed at the interface and by compressing the film it is partly desorbed from the surface.

In actuality during rapid expansion and/or contraction of the surface of a surfactant solution, the surface tension departs from its static value always in such a direction to activate restoring forces which tend to cancel the perturbation. This is the Marangoni effect [4].

Surfactants with long hydrophobe "tails" in sufficiently high concentration promote the formation of bound (highly structured) water. Non-ionic surfactants generally do not change the structure of water because in such compounds the effect of hydrophobe and hydrophyl groups compensate each other.

Since in bound water the molecular diffusivity is lower, by applying surfactants with long hydrophobe "tails" a lower mass transfer coefficient is expected in the surface phase than in presence of non-ionic surfactants [5]. However, because of the higher solubility of oxygen in such systems the oxygen transfer rate can be increased due to the higher driving force in the presence of surface active agents.

d) *Proteins* are surface active; therefore they are adsorbed at the gas/liquid interface. They also reduce the surface tension. Therefore they behave fairly similar to common surfactants. The differences in properties of common surfactants and protein mono-layers play a role in foam formation. Common surfactant monolayers are used as a low viscosity component and protein monolayers as a high viscosity and elasticity compo-nent (stabilizer) in mixtures of foamers [4, 6].

The type of proteins strongly influences the foaming behaviour of culture solutions, e.g., β-casein-like proteins (a flexible, more or less random coil molecule) and lysozyme or bovine serum albumin-like proteins (globular forms) yield different foamability and stability [103, 104].

Surface active agents can have hydrophobic interactions with proteins [90], which again influence the behaviour of the gas/liquid interface.

e) *Glucose* increases the surface tension of water because, similar to inorganic salts, it makes the solution more cohesive and hence it promotes the formation of bound (highly structured) water, in sufficiently high concentrations. Under these conditions excess water molecules and not the solute molecules adsorb at the surface [4, 7].

f) With the model medium no heavy foaming was observed [1]. However, during the cultivation experiments there was considerable *foam formation*.

One can use *antifoam agents* or a mechanical foam destroyer to keep such foam under control.

To select the appropriate antifoam agents the foam properties have to be considered [4, 6].

A foam is thermodynamically unstable, since, once the sheet of liquid is ruptured, it must break into drops with a lower total surface area and therefore with a decrease in the free energy of the system. Foams from pure liquids and gases are highly unstable, but suitable surface active agents can stabilize a foam. The foam stability depends in the early stage on the drainage of liquid from the foam, with consequent weakening of the structure. Drainage is the settling of surplus water to the bottom of the foam, under the action of gravity or surface tension; the latter is due to the lower local pressure in regions of high curvature (at the intersection of three gas bubbles, the so called Plateau border) rather than in the lamella. Hence the liquid is sucked from the lamella into the Plateau border. The film separating the gas bubbles will break when thinned to some critical thickness of the order to 50—100 Å.

Drainage rates can be described by Eq. (2) [4]:

$$\frac{dV}{dt} = \frac{BV_0}{2} (Bt + 1)^{-3/2},$$ (2)

where V is the volume of liquid drainage from the foam, V_0 is the original liquid content, B is a measure of the initial drainage rate (min^{-1}), t is the time (min).
B decreases markedly with increasing expansion and viscosity of the foam. In deriving Eq. (2) it was assumed that the surface viscosity is always so high that the gas/liquid interface is rigid (this assumption appears valid for protein films but not for detergent monolayers).
Hence the constant B will vary directly with v_L^{-1} (v_L = kinematic viscosity of the liquid). Table 1 shows the considerable differences in B for different foams.

Table 1. Drainage of foams after relatively short times [4]

Foaming solution	Expansion of the foam	B of Eq. (2) (min^{-1})
Sodium lauryl sulphate	5	2
Sodium hexadecyl sulphate	9	1
Triton X-100 (non-ionic detergent)	11	0.8
Mearlfoam (protein hydrosylate)	10	0.5

Another effect which influences the stability of foams is the diffusion of air across liquid lamellae due to the pressure difference in the bubbles. The small bubbles will tend to decrease in size and the large bubbles will tend to grow, because the gas pressure is higher in the smaller bubbles than in the larger ones according to $\frac{\sigma}{2r}$.
This effect can be considerable: e.g., in a foam made from Teepol (trade name of Shell International Chemical Co. Ltd. for a mixture of sodium secondary alkyl sulphates and sodium dodecylbenzene sulphonate) solution the number of bubbles decreases to 10% of their initial number after 15 min although no rupture of the liquid films had occured [8]. Since the rate of diffusion of gas through the lamellae varies inversely with their thickness, a quick thinning of the lamellae in foams also accelerates the gas diffusion across the lamellae.
If the film of liquid lies between two charged monolayers consisting, for example, of long-chain sulphate ions, it may resist further thinning after a critical limit is reached. Beyond this further removal of liquid would bring the charged surfaces to close and would set up a high osmotic pressure within the liquid layer by accumulation of counter ions (e.g., Na^+). Although the free energy of the liquid film will always be greater than if it were broken into drops, the energy barrier to local thinning provided by a charged monolayer is often sufficient to stabilize the film. In addition to reducing the draining rate, the surface and bulk viscosities cushion thin liquid films against shocks. The highest foam stability is associated with appreciable surface viscosity and yield value, while solutions yielding foams of very poor stability show very low surface visco-

sity [35]. For protein-stabilized foams the stability and surface viscosity pass through a maximum at the same pH for any given protein [9]. It is significant that the optimum pH is generally close to the isoelectric point of the protein. For foams stabilized with proteins, the surface viscosity and rigidity appear to be the dominant stabilizing factors. It can be shown that very stable foams are formed, if the phases and surfactants form lyotropic liquid crystals [105, 106]. The stability of a liquid film must be greatest if the surface tension strongly resists deforming forces. A shock induced extension of the local area of the lamella causes a surface pressure gradient between the thinned region and the rest of the surface and this, in turn, results in a spreading of molecules from the adjacent parts of the monolayer to the extended region (Marangoni effect). This spreading will stabilize the lamella because the monolayer will carry with them a layer of the adjacent liquid which opposes the thinning due to the shock. The compression modulus C_s^{-1} of the monolayers should, for stability, be as high as possible, since this assists rapid flow into the extended region of the lamella.

Therefore it is recognized that a good foamer is a material that produces a film at the air/water interface which is relatively condensed but which, at the same time, is fluid and capable of rapid changes in π as the surface area varies. These features account for the fact that often the stability of a foam passes through a maximum with the concentration of the foaming agent. The most stable foams are often obtained with mixtures of foamers, one of them being relatively soluble with low surface viscosity and constituting a reservoir of surface active substances for the foam, the other less soluble but with high surface viscosity, thus assuring the cohesiveness of the mixed film.

The foam depressant must be able to replace such a stable film by others not having the required characteristics. This could be achieved by materials forming solid films or by those whose films are completely fluid, such as the polydimethyl siloxanes. In general, the faster the molecules of the foam depressant or breaker can spread on the liquid film, the greater the thinning and the greater the chance of rupture. The higher alkanols are examples of "foam killers" with high spreading coefficients. Antifoam agents generally promote the coalescence of bubbles.

g) Since the *cell* size is much smaller than the bubble size and the density of the cells does not deviate significantly from that of water, their presence does not influence the hydrodynamics of bubble columns as long as the viscosity of the fermentation medium does not change considerably [1]. However, with increasing biomass concentration the viscosity increases and this influences the behaviour of the bubble column [10]. High cell concentration in the foam leads to a high viscosity, with increased foam stability [11, 12]. This latter factor is enhanced by the increased cell concentration in foams compared with the bulk suspension.

h) At pH = 7 and 20 °C the Henry coefficient for CO_2 in pure water is He^{-1} = 0,035 $\frac{moles}{l\,atm}$ [38]. With increasing salt concentration and temperature He^{-1} is reduced [39].

The equilibrium concentration of CO_3^{-2} can be neglected for pH < 7. The ratio CO_2/ HCO_3^- is given at T = 20 °C by [39]:

$$K = \frac{[HCO_3^-]}{[CO_2][OH^-]} = 3 \cdot 10^7$$

e.g., at pH = 7

$$\frac{[HCO_3^-]}{[CO_2]} = 3$$

or at pH = 5

$$\frac{[HCO_3^-]}{[CO_2]} = 0.03$$

The presence of 2.1 g · l^{-1} HCO$_3^-$ (at 20 °C and pH = 7) in comparison with the 10 g · l^{-1} salt solute is not significant with regard to the salt effect.
The main part of the CO$_2$ is desorbed from the liquid. However, with alcohol solutes CO$_2$ is not the only component which is desorbed from the cultivation medium in large amounts, and the desorption of CO$_2$ does not change significantly the dynamic behaviour of the interface.
With glucose solute CO$_2$ is the only ingredient which is desorbed in a large amount. Therefore it can influence the behaviour of the interface, since the mass transfer occurs from a phase of higher kinematic viscosity and lower molecular diffusivity into air with lower kinematic viscosity and higher molecular diffusivity, this can produce interfacial instabilities and an increased mass transfer rate [36, 37].

2.2. Influence of the Properties of the Gas/Liquid Interface on the Behaviour of Bubble Column Reactors

a) Bubble Size and Behaviour

In general three regions can be distinguished in bubble columns:
— an entrance region, at the aerator, where the bubbles are formed;
— a main region;
— an exit region, where the bubbles disengage and the gas leaves the bubbling layer.
In the entrance region the influence of the composition of the fermentation medium on the bubble size depends on the mechanism of bubble formation. The three known mechanisms of bubble formation are:
— At low gas flow rates, separate bubbles are formed. In this "bubbling gas" range the bubble size is controlled by the interfacial and buoyancy forces [1, 13].
— At intermediate gas flow rates, gas jets are formed in a laminar liquid and the bubble size is controlled by the instability of the jet interface [93—95].
— At high gas flow rates, gas jets are formed in a turbulent liquid and the bubble size is controlled by the interfacial forces and the dynamic pressure forces of the local turbulence [96].
Alcohols, surfactants and proteins diminish and salts and glucose raise the surface tension of water and hence the initial bubble size is affected accordingly if the bubbles are formed in the "bubbling gas" range or if their size is controlled by the dynamic pressure of the turbulence.

The initial bubble size is not influenced by the composition of the fermentation medium when the bubbles are formed by the break-up of gas jets due to instability.
However, in several systems the initial bubble size does not play any role [40]:
— in systems with a high bubble coalescence rate, e.g., in pure water or in fermentation media with antifoam agents [45, 48];
— in systems in which the initial bubble size is larger than the dynamical equilibrium bubble size, e.g., in columns with perforated plate aerators and with hole diameters ⩾ 0.5 mm [45, 48];
— in systems with two-component nozzle aerators. In these systems the initial bubble size is extremely small. In the bubble formation region the bubble size is controlled by the dynamical equilibrium bubble size which is determined by the ratio of the turbulent shear stress to the surface tension, that is by the Weber number [Eq. (3)] [1, 14, 40]:

$$We = \frac{\tau d_{B\,max}}{\sigma}, \tag{3}$$

where τ = dynamic pressure of turbulence, $d_{B\,max}$ = maximum diameter of the bubble which is stable at the turbulent shear stress τ.
According to the theory for local isotropic turbulence by Kolmogoroff [15] and to the assumptions of Batchelor [16] this maximal bubble diameter is given by Eq. (4) [67]:

$$d_{B\,max} \cong const. \; \frac{\sigma^{0.6}}{\left(\frac{E}{V}\right)^{0.4} \rho_L^{0.2}}, \tag{4}$$

where $\frac{E}{V}$ is the energy dissipation rate.

Since $d_{B\,max}$ diminishes with decreasing surface tension, the maximum possible bubble diameter $d_{B\,max}$ is smaller in alcohol and surfactant solutions and (slightly) larger in salt and glucose than in pure water. If the initial bubble diameter $d_{B\,in}$ is smaller than $d_{B\,max}$, the acutal bubble diameter d_B lies between these two limits, i.e.,

$$d_{B\,in} \leqslant d_B \leqslant d_{B\,max}.$$

In systems with completely suppressed coalescence $d_{B\,in}$ would be preserved throughout the whole column ($d_{B\,in} \cong d_B$). In systems with a high coalescence rate the dynamical equilibrium bubble diameter dominates, hence $d_B \cong d_{B\,max}$ in every region in the column. In systems with hindered coalescence (alcohol-salt-solutions) and with small initial bubble diameters $d_{B\,in}$ (porous plates, two-component nozzles) the actual bubble size depends on $d_{B\,in}$ and $d_{B\,max}$ as well as on the coalescence rate [40]. For example, in a dilute methanol solution the coalescence rate is moderately hindered and the addition of salt further hinders coalescence. Therefore in a methanol-salt-solution d_B is smaller than that in the corresponding methanol- or salt-solution. This is in contrast with ethanol-solutions where coalescence is strongly hindered, and the addition of salt does not change the coalescence behaviour significantly. However, salt addition increases $d_{B\,in}$ and $d_{B\,max}$ and it follows that d_B in an ethanol solution is smaller than that in an ethanol-salt solution.

In general, in systems with strongly hindered coalescence the addition of salt increases d_B and in systems with moderately hindered coalescence the addition of salt decreases d_B. The addition of glucose to the water has an effect similar to the salt effect; it increases the surface tension (hence also $d_{B\,max}$) and hinders the coalescence. Therefore in glucose-salt solutions $d_{B\,in}$ and $d_{B\,max}$ are large, but because of the hindered coalescence d_B in glucose-salt solution is smaller than that in the corresponding glucose- or salt-solution. The presence of solutes in cultivation medium influences bubble behavior or bubble size as well.

It is well known that in a bubble rising in a pure liquid internal circulation occurs and interfacial area is continuously created at the front stagnation point and continuously destroyed at the rear stagnation point [17].

If the liquid contains surfactants, they are adsorbed on the surface of the bubble at the front stagnation point because on the upstream half of the circulating bubble the surface is newly formed and stretched. The surfactants are then transported to the rear stagnation region by the interfacial flow, where they are accumulated and because of the compression of the interfacial film partly desorbed. By this circulating movement of the interface the surfactants are enriched in the rear of the bubble. A surface tension gradient is formed along the bubble surface with the highest values in the front stagnation region and the lowest values in the rear stagnation region due to the concentration gradients of surfactants on the bubble surface.

The surface pressure $\pi = \sigma_0 - \sigma$ (where σ_0 and σ are the surface tensions of the surfactant free liquid and of the liquid in the interface) is the smallest at the front stagnation point and increases to the rear of the bubble. Hence a "rigid cup" of surfactant film is formed in the rear stagnation region due to the local maximum of the surfactant concentration and the local film compression caused by the interfacial flow. With increasing surfactant concentration or decreasing bubble diameter the relative surface area of bubble covered by this cup increases. Hence the relative surface area of the mobile interface gradually decreases and finally the interfacial movement stops and the bubble behaves as a rigid sphere [18, 19].

If the maximum surface pressure calculated from the force balance exceeds the maximum surface pressure to which the film can be subjected, the adsorbed film will desorb or crumple, or it may rather suddenly be shed to the rear of the bubble as a filament. The bubble may thus lose its surface film. For this reason bubbles larger than some critical radius, may have a calculated maximum surface pressure greater than the adsorbed film can maintain and hence will rise with virtually no retardation [18, 19]. Therefore large bubbles are freely circulating, small bubbles behave as rigid spheres, and intermediate bubbles are partly covered by the rigid cups of water containing surfactants. In the presence of alcohols this behaviour can be changed. The surfactant film can be displaced from the interface by adding sufficient methanol or ethanol to the system. The ability of the solute to displace surfactants depends on its spreading coefficient S, which is defined by [4]:

$$S_{in} = \sigma_{W/A} - (\sigma_{O/A} + \sigma_{O/W}), \tag{5}$$

where $\sigma_{W/A}$, $\sigma_{O/A}$, and $\sigma_{O/W}$ are the surface and/or interfacial tensions between water/

air, oil/air and oil/water (oil refers to the component which is enriched at the interface). In the present publication the "oil" component is soluble in water, i.e., $\sigma_{O/W} = 0$,

and $\quad \sigma_{W/A} = 72.75 \text{ dyn} \cdot \text{cm}^{-1}$ at 20 °C

$\qquad \sigma_{O/A} = 22.6 \quad \text{dyn} \cdot \text{cm}^{-1}$ at 20 °C (methanol)

$\qquad\qquad = 22.3 \quad \text{dyn} \cdot \text{cm}^{-1}$ at 20 °C (ethanol)

hence $\quad S_{in} \;\; = 50.15 \text{ dyn} \cdot \text{cm}^{-1}$ (methanol)

$\qquad\qquad = 50.45 \text{ dyn} \cdot \text{cm}^{-1}$ (ethanol).

Because the spreading coefficients of surfactants are significantly lower than those of alcohols, the latter can partly displace the surfactants from the interface. These short chain alcohols are, however, so readily desorbed from the surface into the bulk liquid (and into the gas) that they depress the circulation of the bubble interfaces much less than the surfactants.

Owing to the decrease in concentration of the alcohols along the co-current flow bubble column reactor, the concentration of the surfactants at the bubble surface increases and the surface tension decreases with increasing distance from the gas entrance. Because of the surface tension gradient the Marangoni effect increases the internal circulation and the velocity of bubble rise and hence also the mass transfer coefficient [20].

With countercurrent operation the concentration of alcohols increases upwards in the column. Therefore the concentration of surfactants at the surface of the bubbles decreases with increasing distance from the gas entrance. The consequent increase in the surface tension upwards in the column causes a reduction of the internal circulation, the velocity of bubble rise and hence the mass transfer coefficient [20].

Therefore, with co-current operation higher mass transfer coefficients, but lower relative gas hold-ups (and specific interfacial area at the same bubble diameter) can be expected, and with counter-current operation lower mass transfer coefficients, but higher gas hold-up (and specific interfacial area at the same bubble diameter) can be expected as compared with the uniform distribution of alcohol which exists in the absence of cell growth.

b) Mass Transfer Coefficient

In unstirred liquids the surfactant monolayers decrease the mass transfer coefficients of O_2 and CO_2 because the polar "head groups" of surface active agents block the mass transfer due to the layer of molecules of bound (highly structured) liquid water which is always present around these "head groups" [5]. This bonded water can appreciably retard the transfer of solute molecules [4]. The highly structured water adjacent to an interface may be eliminated by adding particular salts (e.g., LiCl) or organic compounds (e.g., urea) [21]. For example, the resistance of a monolayer of octanol to the passage of CO_2 was reduced from 300 s cm^{-1} for pure water to 30 s cm^{-1} with 8 M LiCl solution [21]. Under the latter conditions the structured water was apparently almost completely eliminated.

A small amount of methanol has been found to penetrate and somewhat disrupt the film of octadecanol and again the resistance dropped from 300 to about 30 s cm^{-1} though with further increase in the methanol concentration the resistance increased again to about 500 s cm^{-1}, presumably due to the methanol molecules forming mixed micells at higher concentrations and hence promoting the formation of structured water [5].

In dynamic systems the resistance of the monolayer slightly influences the mass transfer resistance at low intensities of turbulence since the overall resistance in the liquid phase is weakly affected by the film. At high intensities of turbulence, however, the monolayer may increase the mass transfer resistance of the liquid considerably by partially damping the eddies of liquid approaching the surface. This damping effect is especially large with protein monolayers [5]. With increasing concentration of protein the interface becomes so resistant to local compression as to be immobile; hence the mass transfer rate is reduced significantly. This is a purely hydrodynamic phenomenon. The effect of protein and other monolayers on mass transfer rates depends quantitatively on the surface compressional modulus C_s^{-1} (reciprocal of the compressibility of the surface film) [22]. With increasing modulus the mass transfer resistance also increases up to a critical value and then remains constant.

Large bubbles oscillate and this increases the mass transfer rate, e.g., the rate of absorption of CO_2 into pure water from large (> 1.5 cm^3) spherical cup bubbles [23] is about 50% higher than the rate calculated from surface renewal theory. This high mass transfer rate occurs when the rear of the bubble is oscillating. Addition of 0.1% hexanol to the water eliminates this rippling, and hence k_L is reduced to the expected value. If instead of n-hexanol a surfactant is added, e.g., 0.01% "Lissapol" (trade name of I.C.I. Ltd. for RO C_2H_4 $(OC_2H_4)_n$OH of mean molecular weight 572), to the water the mass transfer rate is lowered to about 50% of the theoretical rate, since not only the turbulent ripples at the rear of the bubbles are suppressed, but the surface renewal due to circulation of the bubble is also eliminated [23]. This reduction of k_L can be offset by the addition of few percent of short-chain alcohol or acetic acid [24–26].

c) Longitudinal Mixing

The intensity of longitudinal mixing can be strongly influenced by different additives due to the change of the behaviour of the bubbles [49]. Local turbulence is induced at the gas distributor, and this quickly decays if the liquid Reynolds number is not high enough to support the turbulence.

With increasing distance from the aerator the pressure gradually decreases, the gas holdup increases and hence the liquid Reynolds number increases. In liquids with a high coalescence rate some bubbles grow quickly, others remain unchanged; hence the size distribution and the rise velocity of the bubbles become significantly non-uniform. It follows that the two phase flow becomes unstable and a transition into turbulence takes place. In the transition region the intensity of turbulent mixing (backmixing) increases. The admixture of methanol to the medium retards coalescence, stabilizes the system and shifts the transition region to a larger distance from the aerator. The intensity of the mixing remains low up to the transition region. Since coalescence is more strongly suppressed by ethanol than by methanol, the transition region is shifted to an even greater distance from the aerator.

However, by applying n-propanol very small bubbles are formed and their size is largely preserved due to the strong suppression of coalescence. Because of the very low rise velocity of these small bubbles very high gas hold-ups and high linear liquid velocities (Reynolds number) result. This condition can support the turbulence induced at the aerator. As a consequence the intensity of turbulent mixing is high throughout the column [49].

Salt or glucose additives increase the bubble size and decrease the coalescence rate by stabilising the flow; consequently the transition into the turbulente state takes place later than in pure water.

In general additives which diminish the bubble size and/or block the interfacial circulation of the bubbles and hence reduce the rise velocity, increase the gas hold-up and hence the liquid Reynolds number.

On the other hand, additives which suppress coalescence also stabilize the flow; the exception are those solutes which yield such high gas hold-ups that the transition to turbulence takes place at the aerator.

Not only does the coefficient of the turbulent (back) mixing, D_{LB}, depend strongly on the bubbles, but also on the longitudinal (G. I. Taylor) dispersion, D_{Lax}. The latter is evaluated from the distribution of residence times of the liquid elements. The non-uniform bubble size distribution (i.e., non-uniform bubble rise velocity) results in an increase in the longitudinal dispersion coefficient [46, 47].

As can be appreciated from the foregoing survey, the components of the cultivation medium influence the gas/liquid interface, and consequently the bubble size and behaviour. Since the gas hold-up, the relative bubble swarm velocity, the specific interfacial area, the volumetric mass transfer coefficient and the longitudinal mixing in the liquid phase are controlled mainly by the bubble size and bubble behaviour, all of these properties of bubble columns affect not only their design and operation, but also the requirements on the specification of the components and concentrations which make up the cultivation media.

3. Experimental Equipment and Measuring Techniques

Two virtually identical tower reactors were used in the investigations, and both were manufactured by Giovanola Frères.

Apparatus A served the investigations involving different flow conditions with chemical foam control and *Apparatus B* the investigations with mechanical foam control. Apparatus A was equipped with an additional liquid pump allowing both cocurrent and countercurrent operation and apparatus B with a mechanical foam destroyer. The ancillaries of the apparatusses were slightly different. Both units were made up from the following main parts (Fig. 1):

The *columns* had an inner diameter of 150 mm and a height (bubble layer) of 3 150 mm for column A and 2 760 mm for column B made up from stainless steel and glass sections. Three types of aerator were used: a stainless steel *perforated* plate with 141 holes of 0.5 mm diameter and a free surface area of 0.23%, a stainless steel *porous* plate with a mean pore diameter of 17.5 microns and a free surface area of 40%, a two component injector nozzle with a nozzle diameter of 4 mm supplied by M. Zlokarnik, Bayer AG [44].

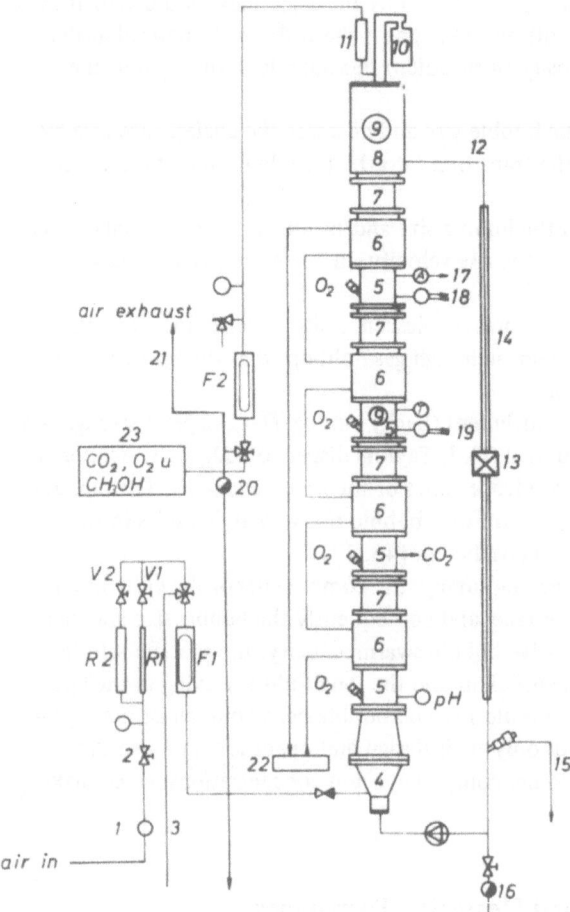

Fig. 1
1 water trap
2 pressure reducing valve
3 steam heating
4 gas distributor
5 stainless steel test sections
6 stainless steel heating/cool-
ing system
7 glass cylinders
8 column head with overflow
9 window
10 foam breaker/destroyer
11 heat exchanger
12 loop connection
13 flow meter
14 loop with heating/cooling
jacket
15 valve for discharge
from the column
16 water trap
17 substrate dosage
18 NH₄OH dosage
19 inoculation valve
20 water trap
21 air exhaust
22 thermostat
23 O₂, CO₂, alcohol measuring
and control instruments
F sterile filters
R flow meter
V needle valve
T resistance thermometer
O₂ oxygen-electrode
CO₂ CO₂-sampling
pH pH-electrode

Oil free air was provided by a gas supply unit and sterilized by filter (F1). The air
entered the tower at its base and was dispersed by aerator (21). In cocurrent operation
the liquid entered the column at the base from a recirculation loop (14). The fermentors
consisted of a bottom section [with the aerator (4)], a head section and column section.
The latter was made up of 12 cylindrical segments:
4 of them (5) consisted of stainless steel tubes and were equipped with 6 fittings for
O₂-electrodes, pH-electrodes, resistance thermometer, CO₂-sampling, for cell density
measurements and for taking samples,
4 of them (6) consisted of stainless steel tubes covered with heating/cooling jackets
and
4 of them (7) consisted of glass cylinders for observing the two-phase flow.
At the half height of column B a plane window (9) was mounted on a segment (5) to
estimate the bubble size distribution by a photographic method [40]. The head of the
column was equipped by an overflow (8) for cocurrent operation, a plane window (9)
to observe the liquid level, a connection (12) to the recirculation loop (14), a gas outlet

with cooler (11), a mechanical foam destroyer (10) on column B [29], two inlets (17) (18) for substrate and NH_4OH-solutions and magnetic valves to regulate their inflow for control of pH- and substrate concentrations. For countercurrent operation in column A the liquid entered the head section from a recirculation loop (14). The loop (14) is covered by a heating/cooling jacket and equipped with an inductive liquid flow meter (13) and a valve (15) for discharging of the liquid. The exhaust gas passed through a gas cooler (11), a sterile filter (F2), a sampling valve and a liquid trap (20). The complete equipment was sterilized by water vapor at 121 °C.

The ancillaries consisted of:

6 polarographic O_2-electrodes with suitable power supply units to measure the longitudinal concentration profiles of O_2 dissolved in the liquid,

a liquid flow meter,

a pH meter with pH-control,

a thermometer with temperature control,

a flame ionisation detector (type R″-5, Ratfisch, München) to analyse the alcohol in the exhaust gas and control the solute feed,

an infrared analyser (Unor, Maihak for column A and Uras 2 T, Hartmann & Braun for Column B) to measure the concentration of CO_2 and a paramagnetic analyser (Oxigor, Maihak for Column A and Oximat 2, Siemens for Column B) to measure the concentration of O_2 in the exhaust gas.

The concentration of CO_2 dissolved in the fermentation medium was measured by a silicon membrane sampling tube coupled to a mass filter (Finnigan, Model 3000) in apparatus A [41], and/or coupled to a four channel mass spectrometer (Varian Mat, Model GD 150/4) in apparatus B [29].

The cell concentration was measured by an Autoanalyser system [42, 43] in apparatus A and/or by a turbidimeter operated by means of reflected light (Fig. 2) [27–29] in apparatus B.

The cell concentration was controlled by gravimetric measurements. Since the response time of the O_2-electrodes was unsatisfactory they were considerably improved [101, 29].

In apparatus B a mechanical foam separator of Giovaniola Frères (Fig. 3) was applied as well as a mechanical foam destroyer (Fig. 4) [29]. The latter was based on a principle invented by the Electrolux Co. and applied in the paper manufacturing industry [30].

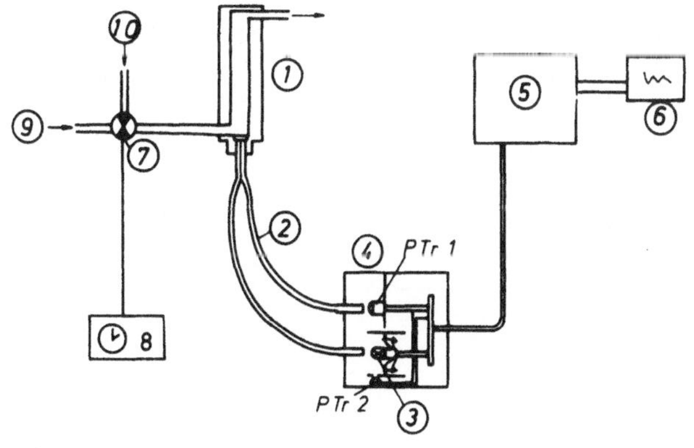

Fig. 2. Turbidimeter
1 measuring cuvette
2 glass fibre cables for primary and reflected light transfer
3 light source
4 light receiver
5 amplifier
6 recorder
7 magnetic valve
8 time switch
9 connection to reactor
10 connection to disinfection solution

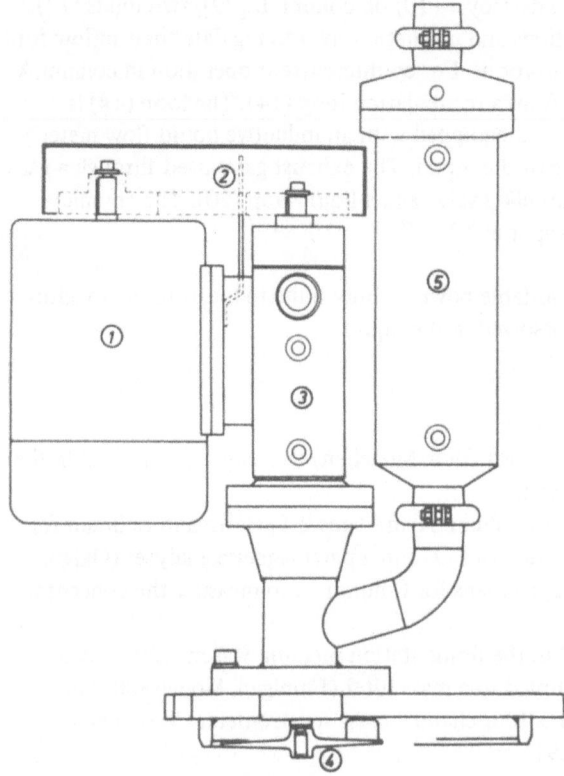

Fig. 3
Mechanical foam breaker
1 variable speed motor
2 transmission
3 cooled shaft
4 rotor
5 gas cooler

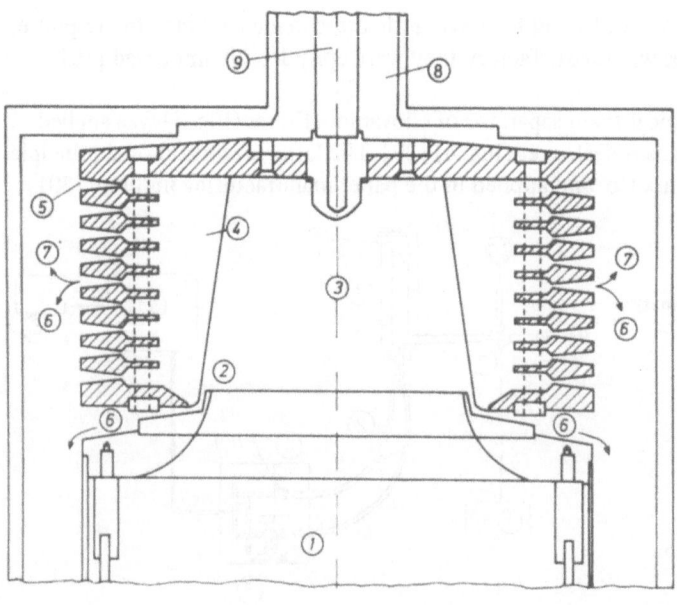

Fig. 4
Mechanical foam
destroyer
1 column head
2 overflow
3 foam
4 baffles
5 venturi nozzles
6 liquid
7 gas
8 gas exit
9 shaft to motor

4. Estimation of Model Parameters

The tower fermentors were characterised by the mean relative gas hold-up ϵ_G, Sauter bubble diameter d_S, specific interfacial area a and/or A, oxygen transfer rate G, volumetric mass transfer coefficient $k_L a$, energy dissipation rate E/V_L and the productivity with regard to the cell mass P_r.

The substrate yield coefficient Y_s, oxygen yield coefficient Y_0 and maximum specific growth rate μ_{max} are characterized by the organism used.

4.1. Definitions and Estimation of the Parameters

The mean relative gas hold-up is defined by Eq. (6)

$$\epsilon_G = \frac{V - V_L}{V} = \frac{H - H_L}{H},\tag{6}$$

where V = volume of the bubbling layer, V_L = volume of the bubble free liquid layer, H = height of the bubbling layer, H_L = height of the bubble free layer.
H and H_L were measured directly on the columns.

The *bubble diameter* distributions were measured at the half height of the column by a photographic method and the photographs were evaluated by means of a particle size analyser (type TZG3, Carl Zeiss Co) [28]. The mean bubble diameter d_B and the Sauter mean diameter d_s were evaluated from the bubble size distribution, the latter according to Eq. (7):

$$d_s = \frac{\sum\limits_{1}^{N} n_i d_i^3}{\sum\limits_{1}^{N} n_i d_i^2},\tag{7}$$

where n_i is the frequency of the bubbles with the diameter d_i.
The specific interfacial area A was calculated from Eq. (8)

$$A = \frac{a'}{V} = \frac{6\,\epsilon_G}{d_s}.\tag{8}$$

The oxygen transfer rate G is defined as

$$G = \frac{Q}{V_L},\tag{9}$$

where

$$Q = \left(\frac{\text{mass O}_2 \text{ transferred and/or consumed}}{\text{time}}\right)$$
$$V_L = \text{ volume of the liquid,}$$

and was calculated from the O_2 balance.

Since the oxygen driving force is the difference between the oxygen concentration at the interface C_L^* and in the bulk liquid C_L, the mass transfer coefficient k_L is defined by Eq. (10).

$$Q = k_L a' (C_L^* - C_L),\tag{10}$$

where C_L^*, C_L are the concentrations of O_2 in the liquid $\left(\dfrac{\text{mass } O_2}{\text{liquid volume}}\right)$

$\quad a'\quad$ gas/liquid interfacial area

$$k_L = \frac{Q}{a'(C_L^* - C_L)}\qquad \left(\frac{\text{volume of liquid}}{\text{interfacial area} \times \text{time}}\right)\tag{10a}$$

or

$$k_L a = \frac{k_L a'}{V_L} = \frac{Q}{V_L(C_L^* - C_L)}\frac{1}{}\left(\frac{1}{\text{time}}\right)\tag{10b}$$

$\quad a\ =\ \dfrac{a'}{V_L}\ $ specific interfacial area.

The relation between a and A is given by Eq. (11):

$$A = a(1 - \epsilon_G).\tag{11}$$

The volumetric mass transfer coefficient $k_L a$ was estimated either from Eq. (10b) using the mass balance on O_2 in the fermentor and the mean driving force $(C_L^* - C_L)$ or by means of the longitudinal concentration profiles for the O_2 in the liquid phase, measured at six positions along the tower. In the latter method a mathematical model was used which is described in Sect. 4.2.

If one assumes that the oxygen concentration in the bulk liquid during O_2-limited growth is negligibly small ($C_L \cong 0$), Eq. (10) simplifies to Eq. (12)

$$Q_{max} = k_L a' C_L^*.\tag{12}$$

The maximum oxygen transfer rate is given by Eq. (13).:

$$G_{max} = \frac{Q_{max}}{V_L} = k_L a\, C_L^*.\tag{13}$$

The *energy dissipation rate* E/V_L is the power E required for gas dispersion in liquid of volume V_L, and can be calculated from the pressure drop of the air across the column [1]. The rate of dissipated energy E^* is that required to separate or destroy the foam, but it does not include the friction losses, since the contribution of the latter quickly diminished with increasing size of the unit.

The productivity is defined as the cell production rate:

$$Pr = \frac{dx}{dt},$$ (14)

where x is the cell concentration, measured as dry cell mass (BTM).
The maximum specific growth rate is defined for an unlimited batch culture by Eq. (15):

$$\mu_{max} = \frac{1}{x}\frac{dx}{dt}.$$ (15)

Pr and μ_{max} can be estimated directly from the measured values of the concentration of dry cell mass vs. time.
The substrate and oxygen yield coefficients are defined by Eqs. (16) and (17):

$$Y_s = \left(\frac{\text{mass cell produced}}{\text{mass substrate consumed}}\right)$$ (16)

$$Y_0 = \left(\frac{\text{mass cell produced}}{\text{mass oxygen consumed}}\right).$$ (17)

4.2. Determination of Liquid Phase Mass Transfer Coefficients

During the growth experiments the oxygen concentration in the liquid phase was measured at six different positions in the bubble column by applying polarographic electrodes at each measuring point. The electrodes were checked with oxygen-free and oxygen saturated culture medium after sterilization. As the experiments were carried out with a continuous feed of substrate the environmental conditions were nearly constant [87] and consequently the variation of oxygen concentration with time during the determination of a single profile was negligible. In order to avoid the inaccuracies involved in the calculation of liquid phase mass transfer coefficients based on simple models like the plug-flow tube reactor (PFTR) or continuous stirred tank reactors (CSTR), a more realistic model was used [58–62] which considers partial axial back-mixing, convection and reaction in the liquid phase as well as mass transfer between both phases. In the gas phase plug-flow was assumed. The oxygen mass balance in the liquid phase (Fig. 5) yields:

$$\frac{1-\epsilon_G}{Bo}\frac{d^2C_L}{dz^2} - \frac{dC_L}{dz} + St\,(C_L^* - C_L) - r\frac{(1-\epsilon_G)\cdot H}{w_{SL}} = 0$$ (18)

with

$$Bo = \frac{w_{SL}\cdot H}{D_{LB}}$$ (19)

and

$$St = \frac{k_L a\cdot H}{w_{SL}}.$$ (20)

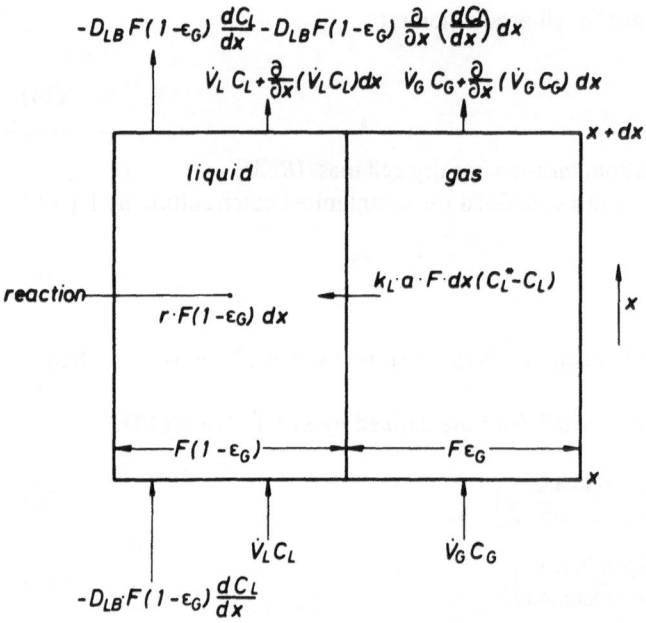

Fig. 5. Volume element for oxygen mass balance in liquid and gas phases

Rewriting Eq. (18) using Henry's law and assuming a linear pressure profile [61]:

$$\frac{1 - \epsilon_G}{Bo} \frac{d^2 x_{Lo}}{dz^2} - \frac{dx_{Lo}}{dz} - St\left\{x_{Lo} - \frac{p^0}{He}[1 + \alpha(1 - z)]x_{Go}\right\} - r\frac{(1 - \epsilon_G) \cdot H}{w_{SL} \sum_j \frac{\rho_j}{M_j}} = 0 \quad (21)$$

with α as the quotient of the hydrostatic to the head pressure in the bubble column:

$$\alpha = \frac{\rho_L \cdot g(1 - \epsilon_G) \cdot H}{p^0}. \quad (22)$$

Since the actual oxygen concentration in the gas phase, x_{Go}, is not known, an additional equation from an oxygen mass balance in the gas phase (Fig. 5) is necessary:

$$\frac{d}{dz}\left\{\bar{w}_{SG} \cdot [1 + \alpha(1 - z)]x_{Go}\right\} - St\frac{RT \sum_j \frac{\rho_j}{M_j}}{p^0}\left\{x_{Lo} - \frac{p^0}{He}[1 + \alpha(1 - z)]x_{Go}\right\} = 0 \quad (23)$$

with $\bar{w}_{SG} = \frac{w_{SG}}{w_{SL}}$ (dimensionless superficial gas velocity).

Owing to decreasing pressure gas expands and the gas velocity has to be considered variable. A total gas phase balance (Fig. 6) gives:

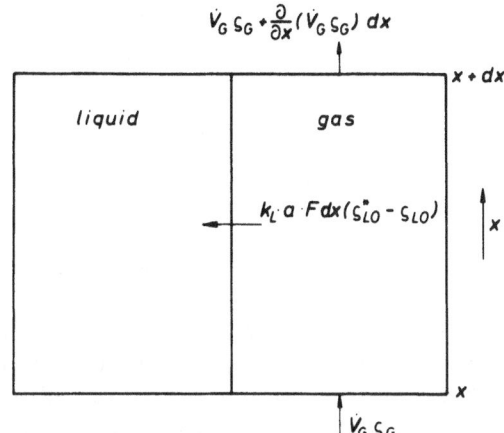

$$\dot{V}_G \, \varsigma_G + \frac{\partial}{\partial x} (\dot{V}_G \, \varsigma_G) \, dx$$

$x + dx$

| liquid | gas |

$k_L \cdot a \cdot F dx (\varsigma_{LO}^* - \varsigma_{LO})$ x

Fig. 6. Volume element for total
mass balance in gas phase

x

$\dot{V}_G \, \varsigma_G$

$$\frac{d\bar{w}_{SG}}{dz} - \frac{1}{1 + \alpha(1 - z)} \left(\alpha \cdot \bar{w}_{SG} + St \cdot \frac{RT \sum_j \frac{\rho_j}{M_j}}{p^0} \left\{ x_{Lo} - \frac{p^0}{He} \left[1 + \alpha(1 - z) \right] x_{Go} \right\} \right) = 0. \quad (24)$$

Equations (21), (23), and (24) represent a system of three differential equations which have to be solved simultaneously. Equation (23) is non-linear. The boundary conditions are: for Eq. (21)

$$x_{Lo}(o) = X_{LoE} + \frac{1 - \epsilon_G}{Bo} \frac{dx_{Lo}(o)}{dz} \tag{25}$$

$$\frac{d x_{Lo}(1)}{dz} = - \frac{r H}{w_{LR} \sum_j \frac{\rho_j}{M_j}}. \tag{26}$$

The oxygen consumption rate r was assumed to be independent from column height Z: For Eq. (23)

$$x_{Go}(o) = x_{GoE} \tag{27}$$

for Eq. (24)

$$\bar{w}_{SG}(o) = w_{SGE}. \tag{28}$$

When attempting to solve this system of differential equations using common numerical methods difficulties appeared due to Eq. (21) whose homogeneous part has Eigenvalues with different signs and absolute values of about 4—17. Therefore a method was used which splits the Laplacian of Eq. (21) into two differential operators [54]. Following this it was possible to calculate oxygen concentration profiles in the liquid phase with given parameters Bo and St. The volumetric mass transfer coefficients $k_L a$ and back-mixing coefficients D_{LB} were then found by varying Bo and St until the quadratic deviation between calculated and measured concentration profiles became a minimum. A heuristic optimum seeking method was used for this stage.

5. Characterization of Microbial Systems

The microorganism used in these studies was a yeast, *Candida boidinii,* which was isolated for its ability to utilize methanol as its sole carbon and energy source [63]. This methanol assimilating yeast is facultative methylotropic and grows also on substrates such as glucose, ethanol or organic acids but it is unable to grow on the following C_1-compounds: Methane, methylamine, formaldehyde, and formate as sole carbon and energy source. Intensive studies over the past five years have shown that in methanol utilizing yeasts the oxidation of methanol is accomplished by three successive oxidative reactions via formaldehyde and formate to carbon dioxide:

$$CH_3OH \longrightarrow HCHO \longrightarrow HCOOH \longrightarrow CO_2. \qquad (29)$$
$$X\ XH_2 \qquad\qquad Y\ YH_2 \qquad\qquad Z\ ZH_2$$

In this pathway for methanol oxidation by yeasts X, Y, and Z represent the different electron acceptors and in each oxidation step 2 electrons are derived from the C_1-substrate. It has been shown that a NAD linked alcohol dehydrogenase activity occurs in all the methanol utilizing yeasts tested. However, when this enzyme is recovered from *Candida boidinii* it does not catalyze the oxidation of methanol but nevertheless is formed constitutively [64]. In methanol-utilizing yeasts a FAD linked alcohol oxidase was found to be responsible for the oxidation of methanol to formaldehyde [64–66]. This enzyme is strictly dependent on oxygen as an electron acceptor ($X = O_2$)

$$CH_3OH + O_2 \rightarrow HCHO + H_2O_2. \qquad (30)$$

The alcohol oxidase from *Candida boidinii* has a broad substrate specificity for short chain primary aliphatic alcohols; the K_m values for methanol and ethanol were calculated to be 2 mM and 4.5 mM, respectively. This indicates that the affinity of alcohol oxidase for primary aliphatic alcohols decreased with the increasing length of the alkyl chain.

Apart from the alcohol oxidase, a catalase is involved in a peroxidative oxidation of methanol by H_2O_2:

$$CH_3OH + H_2O_2 \xrightarrow{\text{catalase}} HCHO + 2\ H_2O. \qquad (31)$$

Therefore it has been suggested that catalase may also oxidize methanol during the growth of *Candida boidinii* on methanol [67]. The activity of the catalase in *Candida boidinii* depends significantly on the carbon source as shown in Table 2.

Table 2. Catalase activity of *Candida boidinii* grown on different carbon sources [67]

Carbon source	Specific activity of catalase (U/mg protein)	
	Exponential phase	Stationary phase
Glucose	70	310
Ethanol	18	320
Methanol	1450	1430

In experiments on the location of the dissimilatory enzymes of methanol metabolism in yeasts [68–70] it has been shown that the alcohol oxidase and the catalase are localized in microbodies. The typical occurrence of microbodies in yeasts grown on methanol suggests that these microbodies may be peroxisomes. It is generally accepted that peroxisomes oxidize intracellular metabolites without generating biologically useful energy such as ATP or $NADH_2$. From this information it is necessary to take into consideration the fact that the oxidation of methanol to formaldehyde in *Candida boidinii* is not coupled to ATP formation.

In all the methanol utilizing yeasts tested so far, a NAD linked and gluthathione dependent formaldehyde dehydrogenase (Y = NAD) has been identified [71–73]. This enzyme has been isolated, purified and characterized from *Candida boidinii* [74]. Furthermore it has been shown that formaldehyde is also oxidized by alcohol oxidase recovered from *Candida boidinii* [75, 76]. The possible oxidation of formaldehyde to formate by alcohol oxidase in vivo is of considerable importance because of the relationship with the energy yield in the dissimilation of methanol by *Candida boidinii* and the consequent yield of biomass.

The last step in the dissimilation of methanol, the oxidation of formate to CO_2; is catalyzed by a NAD dependent formate dehydrogenase (Z = NAD) with a high substrate specifity for formate [74]. The Michaelis constants K_m were observed to be 13 mM for formate and 0.09 mM for NAD in *Candida boidinii*. The oxidation of formaldehyde to CO_2 does not necessarily involve a two step process catalyzed by the aforementioned formaldehyde and formate dehydrogenases. In obligate methylotropic bacteria a dissimilatory hexulosephosphate cycle for the oxidation of formaldehyde to CO_2 has been found [77–79].

Dissimilatory hexulose-phosphate cycle for the oxidation of formaldehyde. HPS: hexulose-phosphate synthase; HPI: hexulose-phosphate isomerase; GPI: glucose-phosphate isomerase; GPDH: glucose-6-phosphate dehydrogenase; PGD: phosphogluconate dehydrogenase.

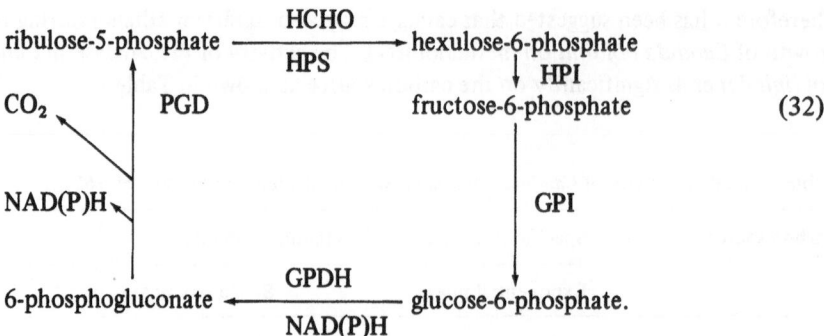

$$HCHO + 2\ NAD(P)^+ + H_2O \rightarrow CO_2 + 2\ NAD(P)H + 2\ H^+. \tag{33}$$

The dissimilatory hexulose phosphate cycle has been found in *Candida boidinii* [80] and may contribute to the oxidation of formaldehyde. Also present in *Candida boidinii* are glucose-6-phosphate isomerase, glucose-6-phosphate dehydrogenase and 6-phosphogluconate dehydrogenase as shown in Table 3. Unfortunately the contribution of the dissimilatory hexulose-phosphate cycle to formaldehyde oxidation in vivo has not yet been quantified.

Table 3. Specific activities of glucose-6-phosphate isomerase (GPI), glucose-6-phosphate dehydrogenase (GPDH) and 6-phosphogluconate dehydrogenase (PGD) in cellfree extracts of *Candida boidinii* grown on different carbon sources [80]

Carbon source	Specific activites $\mu mol \cdot min^{-1} \cdot (mp - protein)^{-1}$		
	GPI	GPDH	PGD
Glucose	1.20	0.60	0.25
Ethanol	0.64	0.25	0.10
Methanol	0.38	0.25	0.12

In methanol utilizing microorganisms there are now three different pathways known in detail which effect the formations of C_3-compounds for the synthesis of cell material from the C_1 units methanol and CO_2: the Calvin Cycle, the ribulose monophosphate cycle and the serine pathway [81]. The overall energy requirements for the synthesis of one molecule of pyruvate from three molecules of methanol by the different assimilation pathways of C_1-units are given below. To enable a comparison between the different pathways, it is assumed that cells can produce two molecules $NADH_2$ by the dissimilation of methanol to CO_2 and that 3 ATP are produced by the oxidation of 1 $NADH_2$ to NAD and 2 ATP by the oxidation of $FADH_2$ to FAD.

1) Calvin cycle (Ribulose diphosphate cycle):

$$3\ CO_2 + 4\ ATP + 5\ H_2O \rightarrow pyruvate + ADP + 4\ P_i. \tag{34}$$

2) Ribulose monophosphate cycle:
2a: Fructose diphosphate aldolase

$$3 \text{ HCHO} + \text{ADP} + \text{P}_i + \text{NAD}^+ \rightarrow \text{pyruvate} + \text{ATP} + \text{NADH}_2. \tag{35}$$

2b: Over Entner Doudoroff pathway

$$3 \text{ HCHO} + \text{NADP}^+ \rightarrow \text{Pyruvate} + \text{NADPH}_2. \tag{36}$$

3) Serine pathway

$$2 \text{ HCHO} + \text{CO}_2 \rightarrow \text{pyruvate}. \tag{37}$$

It is therefore not surprising that among the three different assimilation pathways found in C_1-utilizers, the Calvin Cycle is energetically most unfavourable and that the ribulose monophosphate pathways of formaldehyde fixation are more economical than the serine pathway. Evidence has been obtained that the ribulose monophosphate pathway of formaldehyde fixation is involved in the assimilation of methanol in both *Candida boidinii* [82] and in Candida N-16 [83].

On the basis of these observations theoretical yields of the methanol utilizing yeasts can be calculated by assuming different efficiencies of energy generation from the substrate [84] as shown in Table 4.

Table 4. Theoretical yields of yeasts grown on methanol assuming different efficiencies of formaldehyde oxidation. X, Y, and Z represent arbitrarily chosen symbols for the ATP yield of the oxidation of methanol, formaldehyde and formate, respectively [84]

X	Y	Z	Yield[a] (g cells/g methanol)
0	3	3	0.53
0	2	2	0.47
0	1	2	0.41
0	0	2	0.33

[a] Based on the assumption that methanol is assimilated with the same ATP requirement as in the ribulose monophosphate cycle (fructose diphosphate aldolase variant) for formaldehyde fixation.

In these calculations of the theoretical maximal yield of 0.53 it was assumed that the oxidation of methanol to formaldehyde, catalyzed by alcohol oxidase, does not yield energy. A possible involvement of alcohol oxidase in the in vivo oxidation of formaldehyde may result in the formation of less than 4 moles ATP per mole of formaldehyde oxidized. In Sect. 6 it is shown that the experimental cell yield of *Candida boidinii* is 0.42. This value is close to the assumption that in *Candida boidinii* the oxidation of formaldehyde via formate to CO_2 is coupled to ATP production, and that the dual substrate specifity of alcohol oxidase results in a decreased cell yield. The cell yield of *Candida boidinii* depends on the methanol concentration; increasing the methanol concentration in the culture medium results in a decreased cell yield and an increase in CO_2 production [85, 86]. It seems that the lower cell yields at higher methanol concentrations are due to uncoupling of methanol oxidation and ATP production [85]. When *Candida boidinii* was grown in an extended culture under different concentrations of methanol an increase in the growth rate up to 0.2% (w/v) methanol was observed, followed by a decline due to substrate inhibition [87]. These experiments led to values of K_s of 0.55 g methanol/liter and a maximal growth rate of 0.2 h^{-1}. Unfortunately methanol concentrations consistent with the maximum specific growth rate failed to result in maximal cell yields.

Cultivation conditions. The culture *Candida boidinii* was maintained on Bacto-agar slants as described by Sahm and Wagner [63]. The yeast was transferred to shake flasks containing 1% (w/v) methanol to adapt the culture to growth with methanol as the primary carbon source. After 42 h incubation at 28 °C, 5% of these shake flask cultures were used to inoculate á 10 l reactor as the second subculture. The medium used for preparation of the subcultures, originally described by Sahm and Wagner [63], was supplemented with 1 g/l fructose. The reactor (Type b10, Giovanola Frères, SA, Switzerland) operated with a working volume of 10 l. The temperature during cultivation was maintained at 28 °C. The aeration rate was 1 vvm and the agitator speed 500 rpm. During the cultivation the pH was maintained at 5.0 controlled addition of 5% (w/v) NH_3 solution. The reactor was equipped with a polarographic oxygen electrode for measurement of dissolved oxygen. At the end of the exponential growth phase the culture was used to inoculate the tower bioreactor. The medium used in the tower reactor experiments was described by Reuss *et al.* [87].

6. Growth Kinetics under Various Operation Conditions

To characterize the systems in which growth rates were determined the following operational and process variables were measured:

Operational (Controlled) Variables

- Gas flow rate (superficial gas velocity w_{SG} or volume gas/volume liquid × min, vvm)
- temperature
- pH-value
- substrate concentration C_{gs} in the exhaust gas

Process (State) Variables

- liquid flow rate (superficial liquid velocity w_{SL})
- cell concentration X
- O_2 concentration C_L along the column ⎫ in the fermentation medium
- CO_2 concentration C_{LC} ⎭
- O_2 concentration C_{go} ⎫ in the exhaust gas.
- CO_2 concentration C_{gc} ⎭

In an attempt to understand the operation of the tower reactor for biomass production the influence of the following factors was investigated: the aerator type, the flow conditions, the presence of antifoam agents and/or different substrates. Tower performance was quantified in terms of growth rate μ_m, substrate and oxygen yield coefficients Y_s and Y_o, productivity Pr, oxygen transfer rate G, specific interfacial area $\frac{a'}{V} = A$, volumetric mass transfer coefficient $(k_L a)$ and specific power requirement $\frac{E}{V_L}$.

6.1. Influence of Aerator Type and Flow Conditions

The aerator type has been shown to exert a significant influence on the bubble size, specific interfacial area and volumetric mass transfer coefficient in cell free model media [1]. Therefore experiments were carried out with the methanol/yeast system using a perforated plate, a porous plate and injector nozzles.
The influence of the perforated and porous plates on the cell growth using ethanol and glucose as substrates is discussed in Sect. 6.3.
The substrate concentration and the pH were kept constant and the cell growth was followed by means of the cell density, oxygen consumption and CO_2 production.
In Fig. 7 the (dry) biomass concentration of *Candida boidinii* grown on 0.4% methanol is plotted as function of time for a perforated plate with 0.9 vvm and a porous plate with 0.3–0.8 vvm and 0.8–1.1 vvm. At low cell concentrations the time courses of these growth curves are identical and correspond to unlimited exponential growth with specific growth rate μ_{max}. With increasing cell concentration the growth curve for the

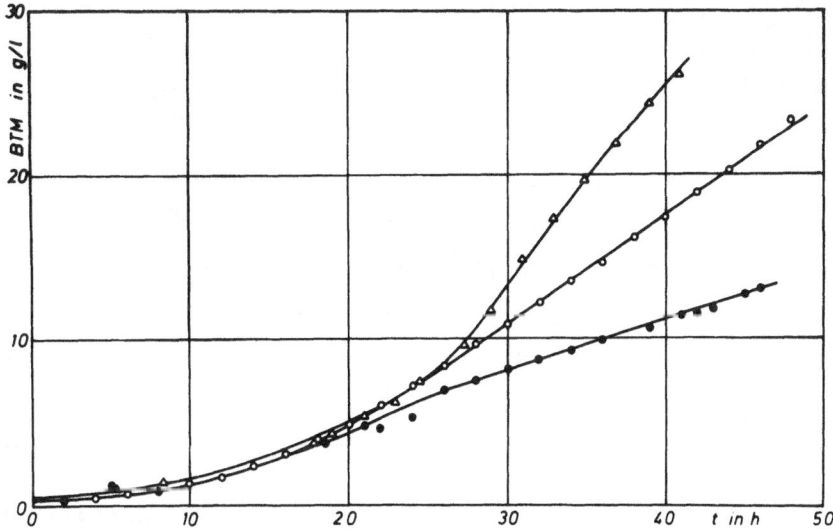

Fig. 7. Growth curve. *Candida boidinii* in the 0.4% methanol with various aerators
● perforated plate w_{SG} = 3.81 cm/s (0.9 vvm); ○ porous plate w_{SG} = 1.9–3.81 cm/s (0.3–0.8 vvm); △ porous plate w_{SG} = 3.6–5.0 cm/s (0.8–1.1 vvm)

perforated plate deviates from the unlimited exponential growth curve and changes into oxygen transfer limited linear growth. With the porous plate this transition occurs at a higher cell concentration and with an increased aeration rate the transition is shifted to even higher cell concentrations.

The slope of the curve in the linear range is given by

$$\frac{dx}{dt} \cong k_L a \left(C_L^* - C_L\right) Y_o. \tag{38}$$

Since Y_o is constant and the variations of both the oxygen partial pressure in the gas phase and of C_L^* were not very different in the three experiments, the different slopes $\frac{dx}{dt}$ are mainly due to the different volumetric mass transfer coefficients ($k_L a$), always providing that the oxygen concentration in the bulk phase was very low. Under oxygen transfer limiting conditions all the process variables respond to a change of ($k_L a$) and/or C_L^*:

In Figs. 8 and 9 experimental data are plotted for a porous plate aerator using various aeration rates.

Curve No. 1 represents the (dry) biomass concentration, curve No. 2 the oxygen consumption and No. 3 the CO_2-production. After the transition from unlimited exponential growth into mass transfer limited growth the slopes of the curves are constant. Sub-

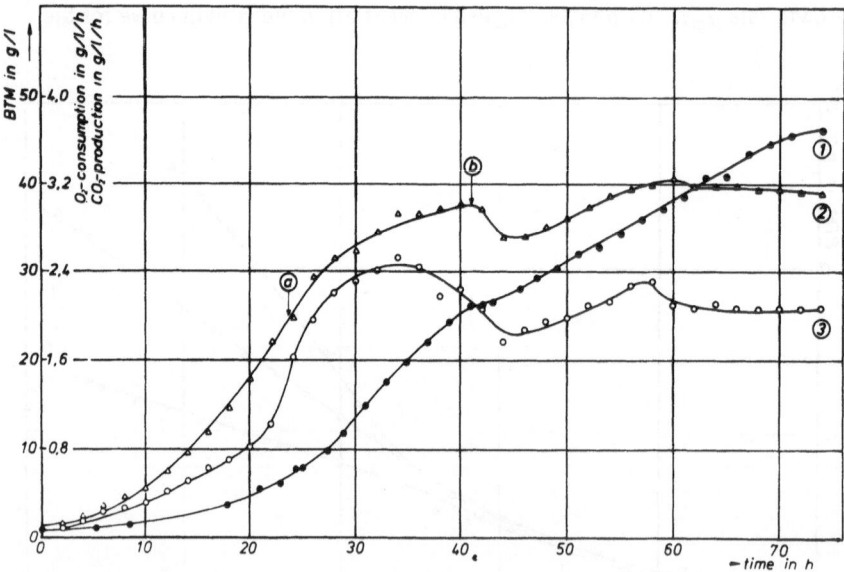

Fig. 8. Growth curve of *Candida boidirii* on 0.4% methanol. Porous plate w_{SG} = 3.6–5.0 cm/s (0.8–1.1 vvm).

			aeration rates:	
1	(dry) biomass BTM	g/l	up to a	1 vvm
2	O_2 consumption	g/lh	a to b	0.8 vvm
3	CO_2 production	g/lh	from b	1 vvm

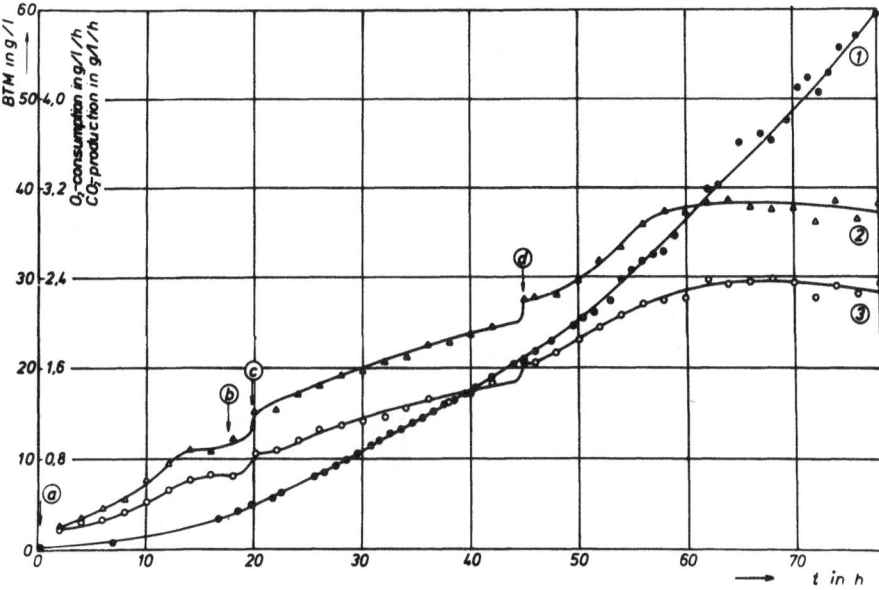

Fig. 9. Growth curve of *Candida boidinii* on 0.4% methanol. Porous plate w_{SG}= 1.9–3.81 cm/s (0.3–0.8 vvm).

			aeration rates:	a	0.3 vvm
1	(dry) biomass BTM	$g \cdot l^{-1}$		b	0.4 vvm
2	O_2 consumption	$g \cdot l^{-1} \cdot h^{-1}$		c	0.8 vvm
3	CO_2 production	$g \cdot l^{-1} \cdot h^{-1}$		d	1.4 vvm

sequently the oxygen consumption and the CO_2 production rates fall, since the growth rate diminished due to the reduced oxygen partial pressure in the gas phase. To compensate this reduction, the aeration rate was increased at the time marked by arrows. Especially in systems with low aeration rate (Fig. 9) the oxygen consumption rate and CO_2 production rate clearly respond to this change.

The longitudinal oxygen concentration profiles are plotted in Figs. 10 and 11. These profiles were measured during the growth of yeast on methanol with perforated and porous plates at various cell concentrations. When using a perforated plate the profiles are essentially flat. With increasing biomass concentration the profiles are nearly parallel to those obtained at lower biomass concentrations. With the porous plate the profiles diminish along the column.

The oxygen concentrations at various axial positions are shown in Fig. 12 as a function of the biomass (BTM) concentration.

In this system an oxygen transfer limitation occurred at BTM concentration of about $16 \ g \cdot l^{-1}$ (Fig. 8) and it follows that a diminution of the relative saturation concentration of oxygen below 20% hinders growth.

The experiments considered above were carried out by co-current flow of air and culture medium. According to Sect. 1 higher specific interfacial areas, but smaller mass transfer coefficients are expected in countercurrent operation compared with co-current

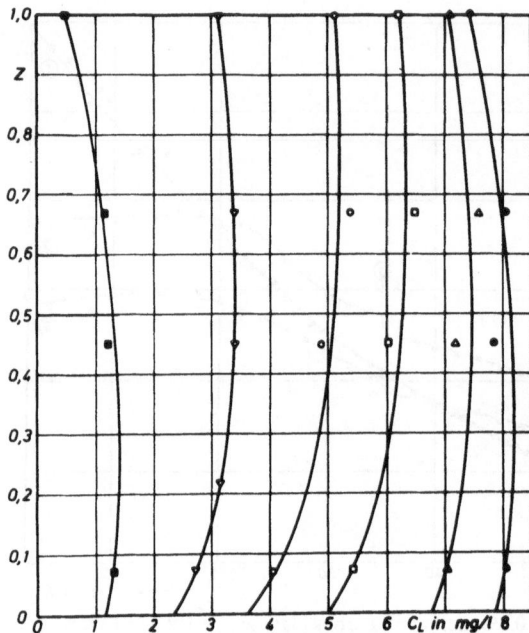

Fig. 10
Longitudinal concentration
profiles of oxygen solute at
various cultivation times and
(dry) biomass (*BTM*) concentrations. *Candida boidinii* with
methanol and perforated plate,
$w_{SG} = 6.32$ cm $\cdot$ s^{-1}. $w_{SL} = 1.74$ cm $\cdot$ s^{-1}

time (h)		BTM (g $\cdot$ 1^{-1})
●	5	0.35
△	10	0.57
□	15	1.05
○	25	4.80
△	30	7.15
■	40	13.05

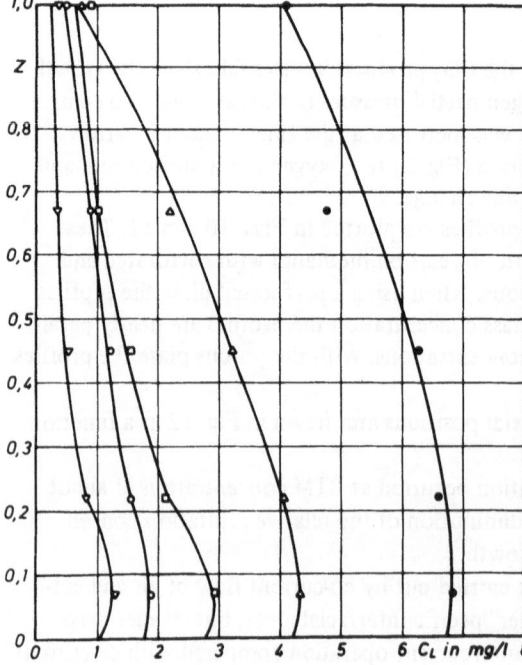

Fig. 11
Longitudinal concentration profiles of
Oxygen solute at various cultivation
times. *Candida boidinii* with methanol.
Porous plate. $w_{SG} = 1.8-6.0$ cm $\cdot$ s^{-1}.
$w_{SL} = 1.3-3.0$ cm/s

time (h)		BTM (g $\cdot$ 1^{-1}
●	8	1.0
△	20	5.0
□	30	11.0
○	45	21.0
▽	78	59.5

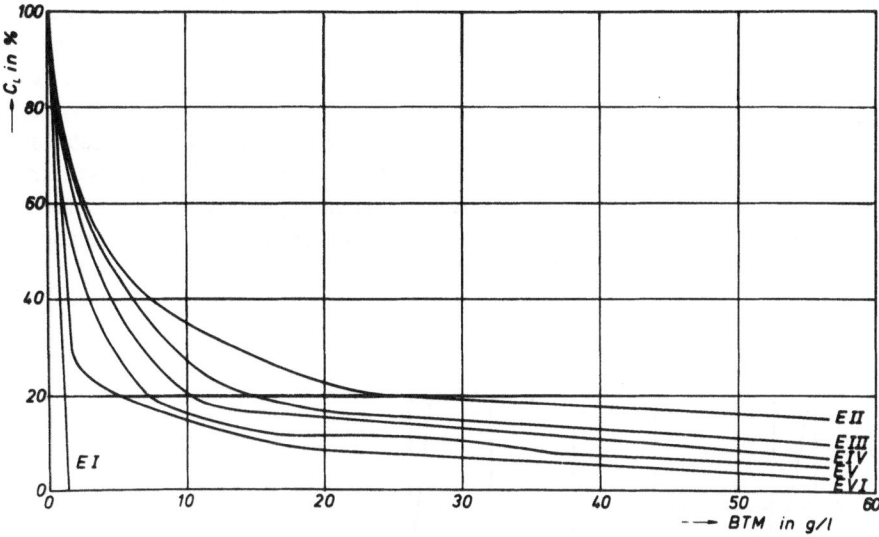

Fig. 12. Relative saturation concentrations of oxygen solute as function of the (dry) biomass concentration at various positions in the column. Methanol substrate and porous plate aerator.

E I: Z = 0.0   E III: 0.22   E V: 0.67
E II: 0.07   E IV: 0.45   E VI: 1.00

operation. However, since the variation of (a) is much larger than that of k_L, the influence of (a) dominates and therefore higher volumetric mass transfer coefficients and oxygen transfer rates are expected for countercurrent rather than co-current operation. Growth curves for *Candida boidinii* on methanol are plotted in Fig. 13 for co-current

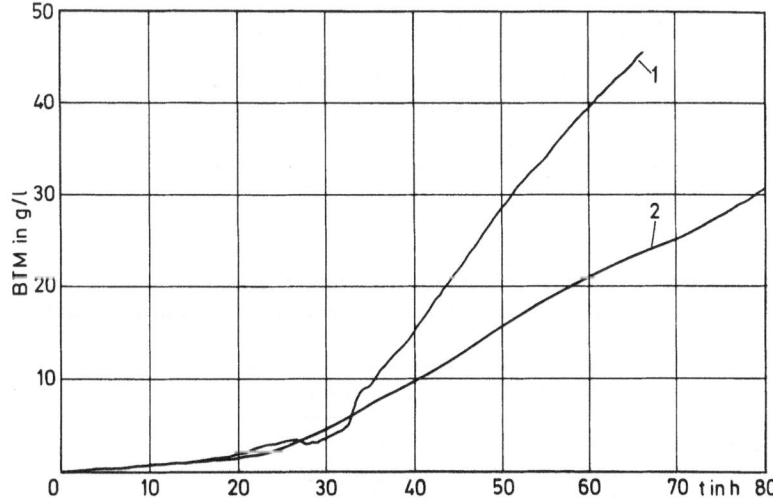

Fig. 13. Comparison of the growth curves in co- and countercurrent operation with methanol substrate in presence of 1 ‰ Ucolub N 115 antifoam agent.
(*1*) counter-current, (*2*) co-current operation

(2) and countercurrent (1) operation. The productivities can be compared directly in the oxygen transfer limited range by means of the slopes of the straight lines. The countercurrent operation yields higher productivity due to a higher oxygen transfer rate. Not only are the volumetric mass transfer coefficients in countercurrent systems higher than those in co-current systems, but the oxygen supply is also favourable due to more uniform longitudinal oxygen concentration profiles (Fig. 14). In co-current systems the volumetric mass transfer coefficient increases proportionally with aeration rate up to fairly high values [1]. In the countercurrent system an optimum aeration rate occured at $w_{SG} = 3.6$ cm $\cdot$ s^{-1} (Fig. 15).

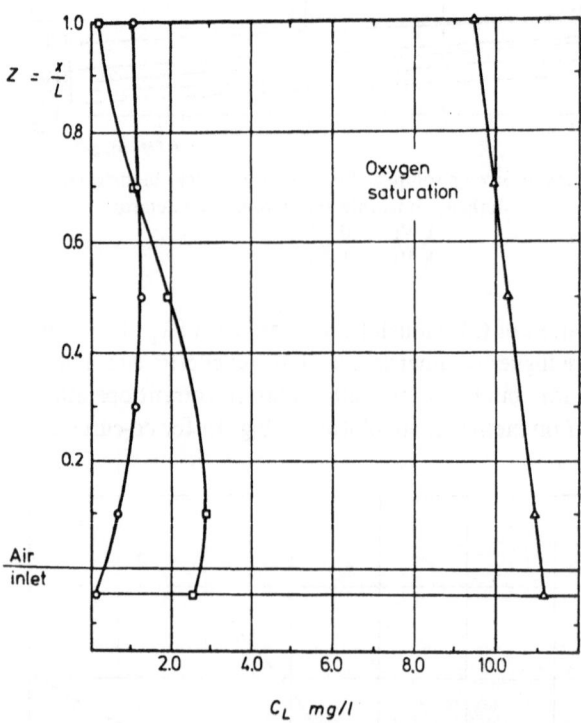

Fig. 14. Comparison of the longitudinal oxygen concentration profiles in the medium in con- and countercurrent operation with methanol substrate in presence of 0.01% Ucolub N 115 antifoam agent

Symbols	□ Concurrent	o Countercurrent
(dry) biomass	17 g $\cdot$ l^{-1}	17 g $\cdot$ l^{-1}
Pr	0.6 g $\cdot$ l$^{-1}\cdot$ h^{-1}	1.36 g $\cdot$ l$^{-1}\cdot$ h^{-1}
w_{SG}^*	7.86 cm $\cdot$ s^{-1}	3.7 cm $\cdot$ s^{-1}
w_{SL}	1.57 cm $\cdot$ s^{-1}	$-$ 7.0 cm $\cdot$ s^{-1}

△ oxygen saturation

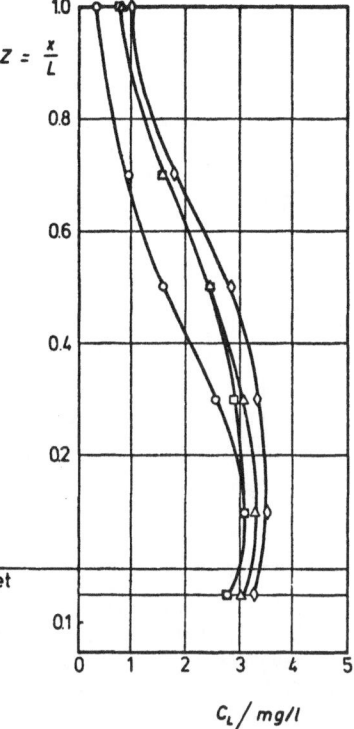

Fig. 15. Comparison of the longitudinal oxygen concentration profiles in countercurrent operation with methanol substrate in presence of 0.01% Ucolub N 115 antifoam agent for various aeration rates w_{SG}

Symbols		○	△	◇	□
w_{SG}^*	$cm \cdot s^{-1}$	5.5	3.8	3.6	3.1
w_{SL}	$cm \cdot s^{-1}$			−7.0	
G_{O_2}	$\dfrac{mg}{l \cdot g}$	2000	2700	2700	2200
$k_L a$	$\dfrac{1}{n}$	244	369	370	308
dry biomass	$g \cdot l^{-1}$	11.0	11.7	12.3	13.0
Pr	$\dfrac{g}{l \cdot h}$	1.36	1.36	1.36	1.36

With increasing gas flow rate the bubbles grow by coalescence. This tendency is higher in countercurrent than in co-current systems because of the longer residence time of the bubbles in the former. The unusually low optimum aeration rate in Fig. 15 is due to the use of antifoam agents, which strongly promote coalescence (see 6.2).

The experiments discussed show the significant influence of the aerator type on the productivity of the tower reactor. In an attempt to explain these large differences, the bubble size distributions were determined before and after inoculation (Fig. 16).

To establish the influence of the different ingredients of the growth medium the mean bubble diameters were obtained for distilled water (a), (A); salt solution (b), (B); salt- and methanol solution (c), (C); for both perforated and porous plates. The influence of the various ingredients on the bubble size agree well with the findings obtained with model media [1].

After inoculation d_B gradually decreased when using a perforated plate (1), but with a porous plate (2) d_B first steeply increased, passed a maximum and then decreased to lower values than those obtained for the perforated plate.

The extent of the initial increase in the bubble diameter immediately after the inoculation in systems containing a porous plate depended on the biological state of the cells used for inoculation e.g., if they were in the exponential growth phase, the increase was small.

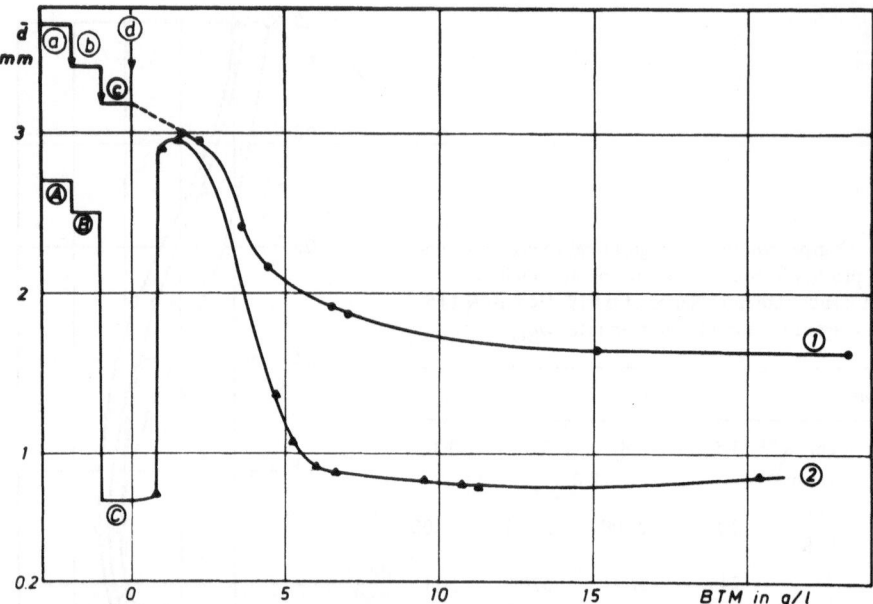

Fig. 16. Mean bubble diameter $\bar{d}_B$ as function of the (dry) biomass concentration. *Candida boidinii* on methanol substrate. Comparison of systems with perforated and porous plate

Perforated plate	Porous plate
a distilled water	*A* distilled water
b 1% salt solution	*B* 1% salt solution
c 0.5% methanol + 1% salt solution	*C* 0.5% methanol + 1% salt solution
d inoculation	*d* inoculation
1 fermentation medium	*2* fermentation medium

At the gas flow rate used ($w_{SG} = 3.1 - 6.4$ cm $\cdot$ s^{-1}) it is probable that no bubbling occurred and since the critical flow rates for the transition from the bubbling into the gas jet state are $w_{SG} = 2.8$ cm $\cdot$ s^{-1} and 8.0 cm $\cdot$ s^{-1}, when calculated using the equations of Ruff [53] and Brauer [52] respectively, it follows that the relation of Kumar and Kuloor [13] cannot be applied for the calculation of the *initial* bubble sizes. A comparison of *initial* bubble sizes calculated according to Kumar and Kuloor [13] and to Rayleigh [94], which is valid in the range investigated [93] with the measured sizes, confirms this supposition (Table 5).

Table 5. Comparison of calculated initial bubble diameter with measured mean bubble diameters in growth systems using methanol ($C_s = 5$ g $\cdot$ l^{-1})

Aerator type	$d_{B\,in}$ (cm) calculated according to		$\bar{d}_B$ (cm) measured at a biomass concentration	
	Kumar [13]	Rayleigh [94]	6 g $\cdot$ l^{-1}	15 g $\cdot$ l^{-1}
Perforated plate	0.31	0.107	0.22	0.14
Porous plate	0.17	0.032	0.087	0,080

If the calculated $d_{B\,in}$ were larger than the measured $\bar{d}_B$, the dynamical equilibrium bubble diameter must prevail. It can be shown, however, using the stability diagram [1] that the bubbles in the growth medium are in the stable range, when using a porous plate aerator. Therefore $\bar{d}_B < d_{B\,max}$ and hence $d_{B\,in} \leqslant \bar{d}_B < d_{B\,max}$. This conclusion excludes the $d_{B\,in}$-values calculated according to Kumar, and confirms the $d_{B\,in}$-values calculated by the Rayleigh relation. Therefore in the methanol growth medium investigated the initial bubble size was always smaller that measured at the half height of the column. Thus the bubbles grow along the column by coalescence and approach the dynamical equilibrium bubble size depending on their tendency for coalescence. This is also true for turbulent flow, in which the initial bubble size is controlled by the local energy dissipation density, since the latter is always higher at the aerator than above it.

Because of the relatively small difference between $d_{B\,in}$ and $d_{B\,max}$ when using a perforated plate, the bubbles attain $d_{B\,max}$ quickly. Therefore, in the main part of these systems $d_B \cong d_{B\,max}$ and hence $\bar{d}_B$ is controlled by the dynamical equilibrium bubble size. With a porous plate $d_{B\,in} \lessdot d_{B\,max}$, therefore the bubbles may not attain dynamical equilibrium in the column. The ratio of the measured mean bubble diameters $\bar{d}_B$ in systems with porous and perforated plates roughly indicates the coalescence tendency of the bubbles. With a ratio of 0.4–0.6 the methanol medium is considered to be a system with a medium coalescence tendency in comparison with the ethanol-salt model medium which had a ratio of 0.2 and pure water with a ratio of 1.0, the former being strongly hindered and the latter having unhindered coalescence.

The specific interfacial areas can be calculated using Eq. (8) and the measured bubble diameters and mean relative gas hold-ups. The resulting data are plotted in Fig. 17 (perforated plate) and Fig. 18 (porous plate) as a function of the dry biomass concentration.

With increasing biomass concentration A increases and approaches $1100\ m^{-1}$ with a perforated plate and $1500\ m^{-1}$ with a porous plate.

To estimate the volumetric mass transfer coefficients $(k_L a)$ the longitudinal concentration profiles of oxygen calculated by means of the model developed in Sect. 4.2 were fitted to the measured data using a hybrid computer [54, 29]. In Figs. 19 and 20 the mole fraction of oxygen in the medium (1), the mole fraction of oxygen in the gas phase (2) and the superficial gas velocity (3) are plotted as a function of the dimensionless column height x/L, where x is the longitudinal distance from the aerator and L is the height of the bubbling layer $H = 276$ cm. In systems with a perforated plate it is possible to fit the calculated longitudinal concentration profiles by the use of a sole Stanton number, $St = k_L a\ \bar{t}$, i.e., with constant volumetric mass transfer coefficient $k_L a$ in the entire column (Fig. 19). In systems with a porous plate a satisfactory fit can only be achieved by assuming a change in the St number along the column. For the sake of simplicity it was assumed that St changes simply with distance along the column. This yields, for the system plotted in Fig. 20, the following relationship for the volumetric mass transfer coefficient

$$(k_L a) = 497 \left(1 - 0.946\,\frac{x}{L}\right) h^{-1}$$

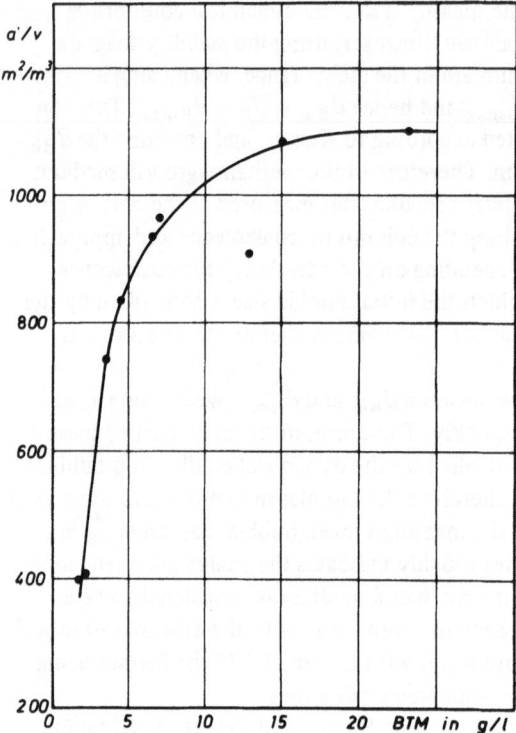

Fig. 17. Specific interfacial area as function of the (dry) biomass concentration. *Candida boidinii* on 0.5% methanol. Perforated plate. $w_{SG} = 4.1-6.4$ cm · s^{-1}

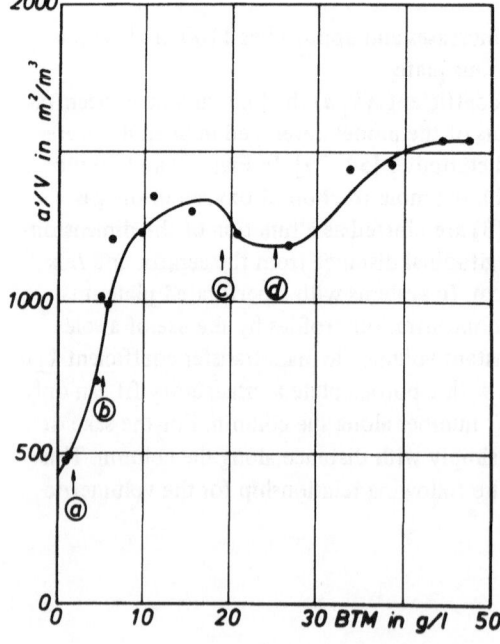

Fig. 18. Specific interfacial area as function of the (dry) biomass concentration. *Candida boidinii* on 0.5% methanol. Porous plate.
$w_{SG} = 3.6-4.8$ cm · s^{-1}
(0.8−1.1 vvm).

a $w_{SG} = 4.8$ cm · s^{-1}
b   3.6 cm · s^{-1}
c   4.1 cm · s^{-1}
d   4.8 cm · s^{-1}

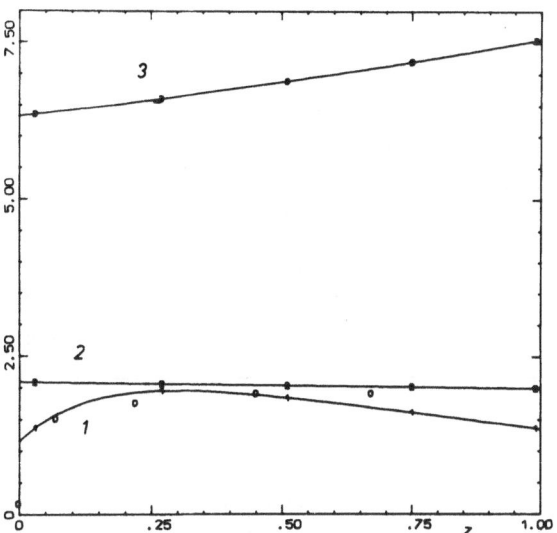

Fig. 19. Fitted longitudinal concentration profiles of oxygen solute.
Candida boidinii on methanol.
Perforated plate. w_{SG} = 6.3 cm
$\cdot$ s^{-1}. w_{SL} = 1.7 cm $\cdot$ s^{-1} $\cdot$
$(k_L a)_{fitted}$ = 294 h^{-1}
1 mole fraction of oxygen
   in the medium $x_{Lo} \cdot 10^6$
2 mole fraction of oxygen
   in the gas phase $x_{LG} \cdot 10$
3 superficial gas velocity
   w_{SG} (cm $\cdot$ s^{-1})

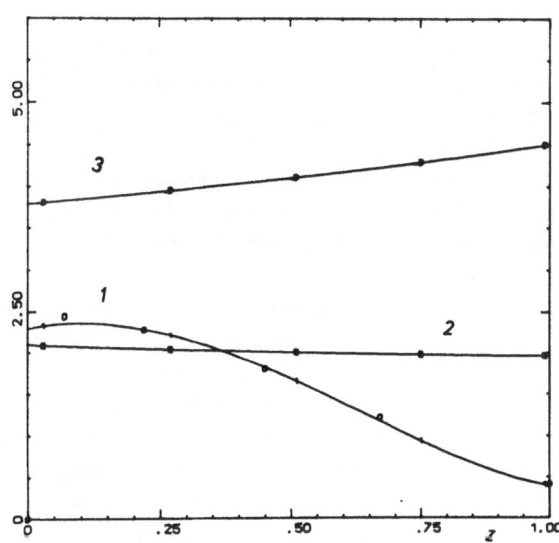

Fig. 20. Fitted longitudinal concentration profiles of oxygen solute.
Candida boidinii on methanol.
Porous plate. w_{SG} = 3.8 cm $\cdot$ s^{-1} $\cdot$
$\cdot$ w_{SL} = 1.9 cm $\cdot$ s^{-1}
$(k_L a)_{fitted}$ = 497 (1−0.946 Z) h^{-1}
1 mole fraction of oxygen
   in the medium $x_{Lo} \cdot 10^6$
2 mole fraction of oxygen
   in the gas phase $x_{LG} \cdot 10$
3 superficial gas velocity
   w_{SG} (cm $\cdot$ s^{-1})

with a Bodenstein number (Bo) of 0.47. The rather crude approximation involved in
this assumption is appreciated and some results indicate [56] that a better fit can be
achieved by a more realistic assumption allowing the largest drop of ($k_L a$) near to the
gas aerator and a relatively slight change in the rest of the column. For these reasons
the ($k_L a$) values presented must be considered as a preliminary first estimate.
A comparison between the volumetric mass transfer coefficients under growth conditions and those in model media is given in Table 6 [91].

Table 6. Comparison of (k_La) values measured in various model media and in the growth medium containing methanol in the absence of antifoam agents ($C_s = 5$ g/l)

Aerator type	Model medium			Growth medium	
	Medium	w_{SG}(cm/s)	(k_La) (s^{-1})	w_{SG} (cm/s)	(k_La) (s^{-1})
Perforated plate	0.5% CH$_3$OH	3.8	0.12	6.3	0.082
	0.5% CH$_3$OH + 1% salt	3.8	0.12		
Porous plate	0.5% CH$_3$OH	3.8	0.38	3.8	0,138[a]
	0.5% CH$_3$OH + 1% salt	2.8	0.48		

[a] Initial value

The volumetric mass transfer coefficients are significantly smaller in the growth medium compared with the model media, though the differences in the corresponding bubble sizes are rather small. This deviation is due to the small mass transfer coefficients k_L existing in the growth media. As discussed in Sect. 2 surface active agents are especially effective in forming rigid cups on the surface of bubbles and diminishing the mass transfer coefficient by suppression or even completely eliminating the free circulation of the bubble surface. In the presence of proteins a monolayer is formed on this rigid surface which causes an additional resistance to the mass transfer. These two effects can result in a change in k_L by a factor of up to ten [5, 25, 55].

In spite of the drop in (k_La) in the growth medium, there still exists a significant influence of the aerator type on the volumetric mass transfer coefficient. This latter effect is due to the different bubble sizes and specific surface areas produced by the various aerators.

Another characteristic difference of the volumetric mass transfer coefficients obtained with perforated and porous plates is the uniform value over the entire column with the former and the drop along the column with the latter. As shown previously [1] in bubble columns with a perforated plate the bubble diameter is mainly controlled by the dynamical equilibrium bubble size and the latter depends only on the energy dissipation rate provided the liquid phase which is completely mixed (constant composition). Since the composition changes only slightly along such columns, the mean bubble diameters, specific surface areas and volumetric mass transfer coefficients are also nearly constant throughout the columns. In columns with a porous plate distributor the bubble size at the distributor is much smaller than the dynamical equilibrium bubble size. Therefore in systems with partially hindered coalescence (methanol medium) the bubble size slowly increases along the column and gradually approaches but does not attain the dynamical equilibrium size. Therefore the bubble size, the specific surface area and the volumetric mass transfer coefficient are dependent on position.

Most of the systems considered up to this point were free of antifoam agents. Since antifoam agents promote the coalescence of bubbles, cultivation media with antifoam agents behave similar to pure liquids, in which the dynamical equilibrium bubble diameter is quickly achieved. Therefore smaller differences in $k_L a$ are expected in antifoam free and antifoam containing systems with a perforated plate than with a porous plate. In systems with a perforated plate in the presence of antifoam agents since the dynamical equilibrium bubble diameter changes only slightly along the column, it is possible to fit the calculated concentration profiles to those measured by using a constant Stanton number. In media with the same composition the same dynamical equilibrium bubble diameter exists for a given energy dissipation rate. Perforated plates with a smaller free surface area yield smaller dynamical equilibrium bubble diameters than those with a larger free surface area due to the higher local energy dissipation rate in the former. Porous plates yield even higher local energy dissipation rates. Therefore higher specific surface areas and volumetric mass transfer coefficients can be obtained in columns with a porous plate of small free surface area than one with a large free surface area and antifoam agents (Table 7).

Two component nozzles have an even higher energy requirement; therefore they yield even higher $k_L a$ than porous plates as shown in [1] and in Table 7.

Table 7. Comparison of the volumetric mass transfer coefficients measured in methanol media with antifoam agents and various gas distributors in co-current and countercurrent operation. *Candida boidinii*, $C_s = 3 \text{ g} \cdot \text{l}^{-1}$

Co-current operation

Aerator type	Perforated plate	Perforated plate	Porous plate with mean pore diameter $\delta_p = 17.5 \ \mu m$	Injector nozzle with nozzle diameter $d_N = 4$ mm
	f.s.a.:[a] 28% $D_\ddot{o} = 0.5$ mm	f.s.a.:[a] 0.23% $D_\ddot{o} = 0.5$ mm		
w_{SG}^* (cm/s)	6.4	6.4	8.4	12.2
$(k_L a)$ (s^{-1})	0.033	0.042	0.060	0.158

Countercurrent operation

Aerator type	Porous plate with mean pore diameter: $\delta_p = 17.5 \ \mu m$
w_{SG}^* (cm/s)	3.7
$(k_L a)$ (s^{-1})	0.108

[a] f.s.a. = free surface area.

Higher volumetric mass transfer coefficients can be achieved in countercurrent operation rather than in the corresponding co-current system.

It can be concluded that the type of gas distributor and flow conditions significantly influence the volumetric mass transfer coefficient ($k_L a$), the oxygen transfer rate and hence the biomass productivity in the oxygen transfer limited range regardless of the presence or absence of antifoam agents (Tables 6–8).

Table 8. Comparison of biomass producitivities measured in the oxygen transfer limited range with *Candida boidinii* grown on methanol with and without antifoam agents in columns with various types of aerator.

In the presence of antifoam agents $C_s = 3 \text{ g} \cdot 1^{-1}$

Aerator type	Perforated plate	Porous plate with a mean	Injector nozzle with nozzle
	f.s.a.: 0.23%	pore diameter	diameter
	$D_ö = 0.5$ mm	$\delta_p = 17.5 \ \mu m$	$d_N = 4$ mm
w_{SG}^* (cm/s)	6.4	8.4	12.2
$P_r\left(\frac{g}{1\,h}\right)$	0.38	0,58	1.3
$\frac{E}{V}\left(\frac{kW}{m^3}\right)$	0.9	1.2	4.2

In the absence of antifoam agents $C_s = 5 \text{ g} \cdot 1^{-1}$

Aerator type	Perforated plate f.s.a.: 0.23% $D_ö = 0.5$ ṁm		Porous plate with mean pore diameter $\delta_p = 17.5 \ \mu m$		
w_{SG} (cm/s)	3.8	6.3	3.8	5.0	6.0
$P_r\left(\frac{g}{1\,h}\right)$	0.31	0.35	0,68	1.24	1.36
$\frac{E}{V}\left(\frac{kW}{m^3}\right)$	0.43	0.71	0.43	0.56	0.67
$\frac{E^*}{V}\left(\frac{kW}{m^3}\right)$	$< 1^a$	$< 1^a$	$< 1^a$	$< 1^a$	$< 1^a$

[a] Depends on the biomass concentration.

6.2. Influence of Antifoam Agents

In Sect. 2 the properties of foams were discussed. It was pointed out that changing the surface tension affects the tendency for foam formation, since the free energy of foam formation is proportional to the product of surface tension and total interfacial area. The surface tension also influences the stability of the foam, since a change in surface tension affects the excess pressure in the foam bubble according to $\frac{\sigma}{2R}$.

The collapse of a stationary foam is mainly influenced by the drainage of the liquid from the lamella (thinning). In practice the foam is exposed to different stresses due to liquid flow or due to the presence of a mechanical foam destroyer, which abruptly deform the foam lamellae. Every deformation which is accompanied by an increase in the surface area of the lamella will disturb the adsorption equilibrium of surface active agents on the interface and reduce the amount adsorbed per unit area. Thus the *surface tension rises locally* and this counteracts further extension of the surface and tends to restore the initial equilibrium in such a way that the local surface pressure gradient induces a spreading of molecules from adjacent parts of the monolayer to the extended region. This spreading stabilizes the lamella because the monolayer drags along the adjacent liquid and so opposes the thinning of the lamella. The compression modulus of the monolayers needs to be high to cause a rapid flow into the extended region. This restoring process can be hindered by *antifoam agents*, e.g., silicones, polymethylsiloxane, tributylphosphate, and polyethylene-propylene-oxide which are strongly adsorbed on the interface without having the necessary high compression modulus to stabilize the foam.

However, it is necessary to use the optimal concentration of antifoam agents, since too low or high concentration can stabilize the foam [92]. Similarly water soluble foam stabilizers can destabilize the foam, if they are applied in too high concentrations.

If the bulk concentration of the surface active substance is high, the initial surface tension is restored in the thinned region by adsorption without transporting liquid to the extended area. Adsorption will therefore not rethicken the liquid film. The lower foam persistence often observed at high detergent concentrations is caused by this effect.

"Foam breakers" e.g., higher alcohols, especially octanol, cause a local rupturing of the liquid film due to *diminished local surface tension,* and the resulting higher surface pressure over a small region. Thus, spreading occurs from this local region and as the film spreads, it carries with it a layer of the underlying liquid, thus thinning the lamella to the critical distance 50–100 Å, where an essentially spontaneous rupture occurs caused by mechanical shock or thermal fluctuation.

Since the coalescence process is also based on a thinning of the liquid film between two "touching" bubbles [57], *foam "breakers"* which hinder the restoring of the thinning, and *anti-foam agents,* which accelerate the thinning, also promote coalescence. This means that the dynamical equilibrium diameter is likely to be maintained in the presence of antifoam agents. However, Eq. (4) cannot be used to estimate $d_{B\,max}$ in fermentation media containing antifoam agents, since the turbulence is significantly damped in the region near to the monolayer of antifoam agents. The turbulent eddies decay near to the interface with the resultant decrease in the dynamic pressure of the turbulence τ; in turn this causes an increase in $d_{B\,max}$ in such a way that the critical We-number [Eq. (3)] remains constant.

The coalescence-promoting and dynamical equilibrium bubble size increasing effect of antifoam agents can clearly be recognized from Fig. 21, where ϵ_G is plotted as a function of the superficial gas velocity for a porous plate and distilled water (1), 1% methanol-solution (2), 1% methanol + 1% salt solution (3), 0.01% Ucolub N 115 (4), 0.1% Ucolub N 115 (5), 0.1% Ucolub N 115 + 1% salt solution (6), 0.1% Ucolub N 115 + 1% methanol + 1% salt solution (7).

The antifoam agent Ucolub N 115 (manufacturer Brenntag) is a water insoluble polyoxyethylene-propylene copolymer with a mean molecular weight of 2500 and a kinematic viscosity of 240 centistokes at 50 °C. Figure 21 indicates that at very low concentrations this antifoam agent has a favourable influence on ϵ_G at low gas flow rates (Curve 4), probably due to its slight surface tension depressing and coalescence hindering effects. Such a phenomenon is not unusual. With increasing concentration and gas flow rate the opposite effects begin to predominate (Curve 5).

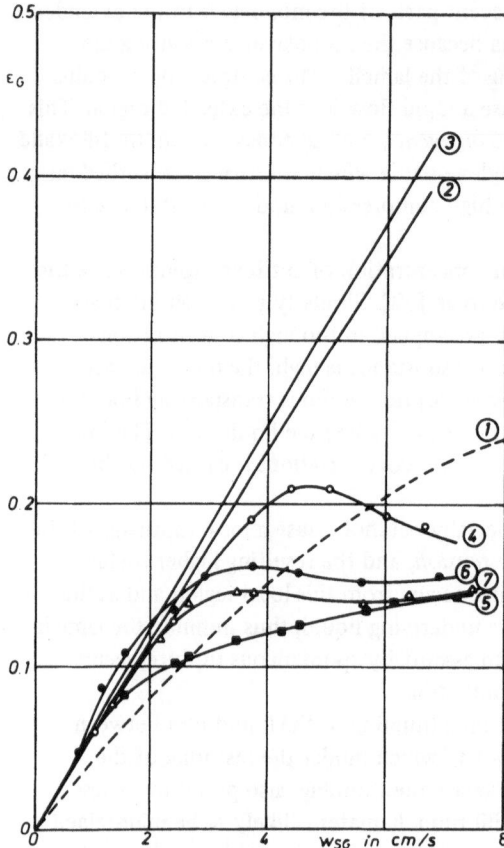

Fig. 21. Influence of the antifoam
agent Ucolub N 115 on the mean relative gas hold up ϵ_G as function of the
superficial gas velocity w_{SG}.

  - - - *1*   distilled water
——— *2*   1% methanol solution
——— *3*   1% methanol + 1% salt
           solution
○   *4*   0.01% Ucolub in water
■   *5*   0.1% Ucolub in water
●   *6*   0.1% Ucolub + 1% salt
           solution
△   *7*   1% salt + 1% methanol +
           0,1% Ucolub solution

A comparison of Curves 2 (1% methanol) and 3 (1% methanol + 1% salt) with Curve 7
(1% methanol + 1% salt + 0.1% Ucolub N 115) indicates the dramatic reduction of ϵ_G,
especially at higher gas velocities. It is interesting to note that the mineral salts medium
yields higher ϵ_G values in the absence of methanol (Curves 6 and 7). It can be deduced
from Fig. 21 that the addition of the antifoam agent causes the bubble size to grow and
the specific surface area and volumetric mass transfer coefficient to be reduced.
This effect can clearly be recognized from Fig. 22. A comparison between the volumetric mass transfer coefficients in methanol based medium both with and without
0.1% Ucolub indicates the severe reduction of $k_L a$ especially at high gas velocities. For
comparison $(k_L a)$ is also plotted for pure water and with 0.01% and 0.1% Ucolub. At
low gas flow rates the latter yields higher $(k_L a)$ than pure water. At the superficial gas
velocities used in the fermentations carried out in the present work, severe deterioration
of the performance can be expected. In Table 9 the volumetric mass transfer coefficients
and the cell mass productivities are compared for fermentations carried out in the
absence and presence of the antifoam agent but at otherwise the same conditions. It is

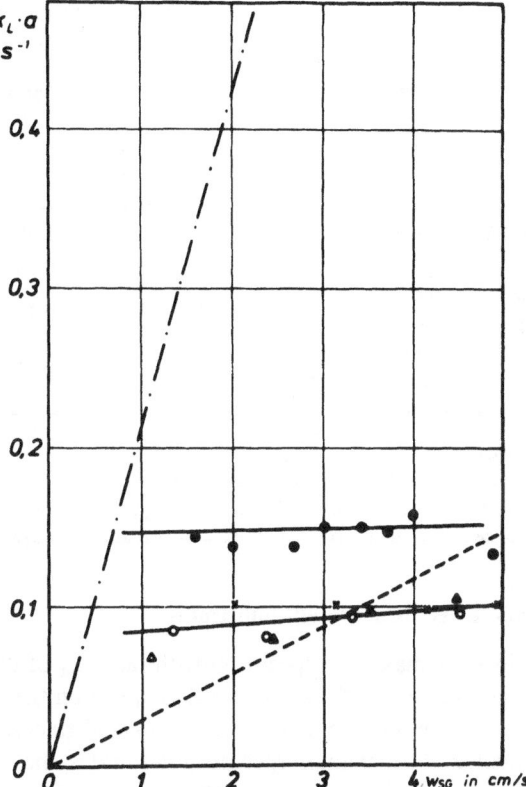

Fig. 22. Influence of the antifoam
agent Ucolub N 115 on the volumetric
mass transfer coefficient ($k_L a$) as function of the superficial gas velocity w_{SG}.

X	0.1% Ucolub in water
△	0.01% Ucolub in water
○	0.1% Ucolub + 1% methanol solution
●	0.1% Ucolub + 1% methanol + 1% salt solution
— — —	distilled water
—·—	1% methanol solution

clear that the performance of the tower reactors is significantly higher when the antifoam agents are absent. However, the alternative of controlling the foam by mechanical
foam breaker or destroyer increases the specific energy requirement of the fermentation.

The foam can be controlled by the mechanical foam breaker of Giovaniola Frères (Fig. 3)
in systems with *Candida boidinii* and methanol as substrate without any difficulty.
However, the energy requirement for foam control is considerable (Table 9).

As discussed in Sect. 6.3 ethanol and glucose were also used as substrates. During the
batch growth on glucose, foam formation was observed and this could be controlled by
the foam breaker.

However, during growth on ethanol, foam with a high mechanical stability is formed.
Since the enrichment of the cells in this foam was particularly high, the foam-cell system
had a fairly high yield stress (plastic rheological behaviour), and in spite of a high energy
input could not be controlled by the foam breaker. To solve the foam control problem
a foam destroyer was developed (Fig. 4), which worked quite satisfactorily, though its
energy requirement is fairly high. The optimal foam control system has not been
developed and systematic investigations are needed to evaluate the chemical and
mechanical properties of the foams and their dynamical behaviour.

Table 9. Comparison of oxygen transfer rate, volumetric mass transfer coefficients and productivities obtained with *Candida boidinii* on methanol in the presence of Ucolub antifoam agent

		Perforated plate			Porous plate			
	Unit	–	–	Ucolub	–	–	–	Ucolub
w_{SG}	(cm/s)	3.8	6.3	6.4[a]	3.8	5.0	6.0	8.4[a]
OTR	mg/lh	820	920	1200	1800	3100	3200	1900
Pr	g/lh	0.31	0.35	0.38	0.68	1.24	1.36	0.58
$\dfrac{E}{V}$	$\dfrac{kW}{m^3}$	0.43	0.71[c]	0.9[c]	0.43	0.56	–	1.2
$\dfrac{E^*}{V}$	$\dfrac{kW}{m^3}$	< 1[b]	< 1[b]	–	< 1[b]	< 1[b]	< 1[b]	–

[a] w_{SG}^*.
[b] Varies with the biomass concentration.
[c] E/V also varies with ϵ_G. At the same w_{SG} in the presence of Ucolub ϵ_G is lower (see Fig. 21) and hence E/V is higher than in absence of this antifoam agent.

6.3. Effect of Different Substrates

Since the maximum specific growth rates μ_m of the yeast used as well as the volumetric mass transfer coefficients strongly depend on the substrate, it is necessary to compare batch growths with methanol, ethanol and glucose. The basis of such a comparison involves the maximum specific growth rates, the substrate and oxygen yield coefficients, the bubble diameters, the specific surface areas, the volumetric mass transfer coefficients, productivities and specific energy requirements.

In Fig. 23 the cell concentration is plotted as a function of the batch time for ethanol using both perforated and porous plates as gas distributors.

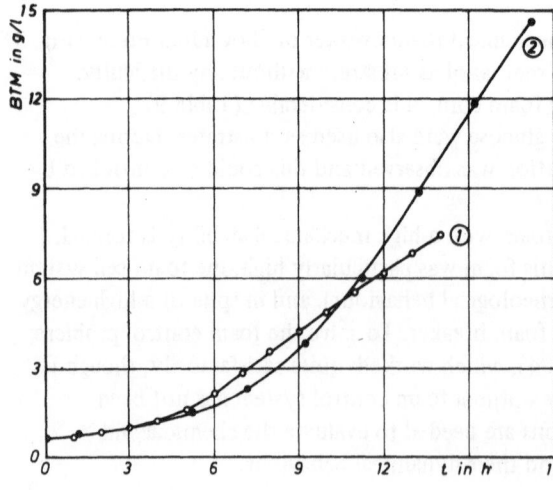

Fig. 23. Growth curves for *Candida boidinii* on 0.4% ethanol. (1) Perforated plate. w_{SG} = 2.98–3.57 cm · s^{-1} (0.63–0.75 vvm). (2) porous plate w_{SG} = 1.7 cm/s (0.5 vvm)

At low cell density, where the growth is not limited by the oxygen supply the two curves coincide. The growth curve obtained when using the perforated plate deviates earlier from the unlimited curve and has a smaller slope in the oxygen transfer limited range when compared with the growth curve for the porous plate (the differences in the growth curves up to 10.5 h are due to the different inoculum used in the two experiments). The different slopes indicate again the great influence of the aerator type (see 6.1).

The highest productivity was obtained with glucose in the unlimited growth region (Fig. 25), because the yeast *Candida boidinii* has its highest specific growth rate, μ_m, with glucose. The lowest productivities were measured when using methanol (Fig. 7) corresponding to the lowest μ_m for the yeast (Table 10). In the oxygen mass transfer limited region the highest productivites were found again with glucose due to the inter-mediate volumetric mass transfer coefficient which exists in glucose media and because of the very high oxygen yield coefficient Y_0 of the cells. The lowest productivities were obtained with methanol, because of the low volumetric mass transfer coefficient and the low oxygen yield coefficient of the cells. Ethanol has an intermediate position because of the intermediate μ_m and Y_0 (Table 10) [29, 91].

Table 10. Comparison of μ_m, Y_s, and Y_0 of *Candida boidinii* for different substrates

Substrate	C_s g/l	μ_m (h^{-1})	Y_s (−)	Y_0 (−)
Methanol	5	0.104	0.42	0.38
Ethanol	5	0.208	0.68	0.66
Glucose	5	0.327	0.48	1.42

The substrate yield coefficient is the highest with ethanol and the lowest with methanol (Table 10). As pointed out in Sect. 5 the theoretical maximum yield coefficient on methanol calculated using the method of Harder [84] is 0.53 g of cell per g methanol. Consideration of the dissimilation pathway discussed in Sect. 5 suggests that the theo-retical cell yield will be between 0.44 and 0.41. These values are in good agreement with the experimental data. The cell yields on ethanol and glucose are comparable with the value published in the literature [88, 89]. The oxygen yield coefficient Y_0 depends on Y_s, as Fig. 24 indicates. These relations were calculated from the O, C, H, and N balances of the cell [97]. Though the cells have a lower oxygen yield coefficient with ethanol than with methanol at the same substrate yield coefficient their oxygen yield coefficient is, in practice, higher with ethanol than with methanol because of the high ethanol yield coefficient. The actual substrate and oxygen yield coefficients for *Candida boidinii* are indicated in Fig. 24 by arrows. The concentrations of biomass, O_2 and CO_2 are plotted in Figs. 25 and 26 as a function of time for ethanol and glucose sub-strates. A comparison of Figs. 8, 9, 26, and 27 indicates that the limitation of growth by oxygen transfer occurs at lower biomass concentrations for ethanol than for glucose. These limits are also plotted as a function of the *OTR* in Fig. 28.

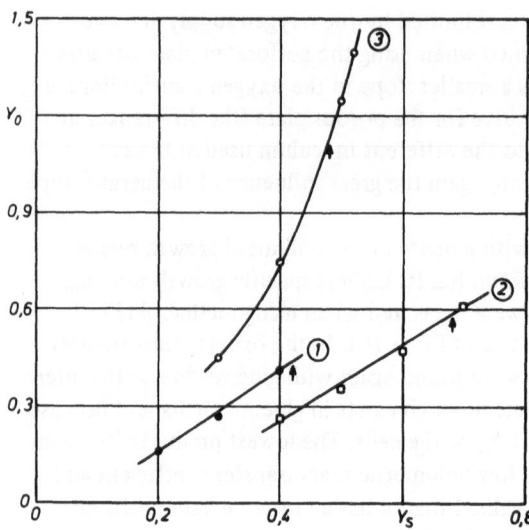

Fig. 24. Calculated and measured oxygen yield coefficients as function of the substrate yield coefficient *Candida boidinii* on (1) methanol (2) ethanol and (3) glucose.
31% O, 47% C, 7.5% N and 6.5 H

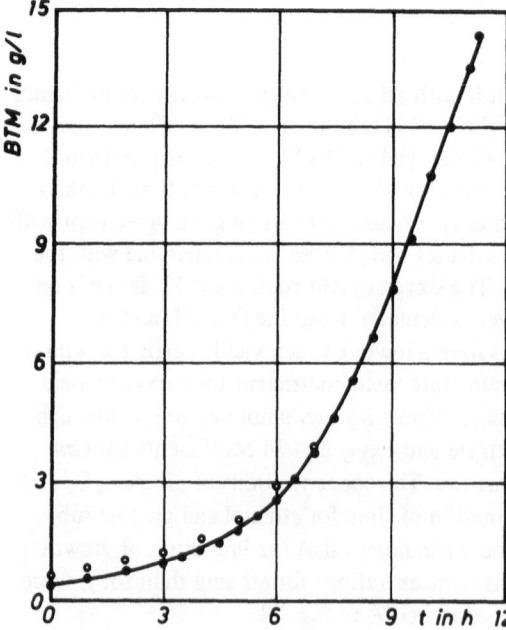

Fig. 25. Growth curve of *Candida boidinii* on glucose. ● Porous plate, w_{SG} = 2.86–3.34 cm/s (0.60–0.71 vvm), ○ porous plate (batch operation with 1% glucose) w_{SG} = 3.10– 3.57 cm · s⁻¹ (0.65–0.75 vvm)

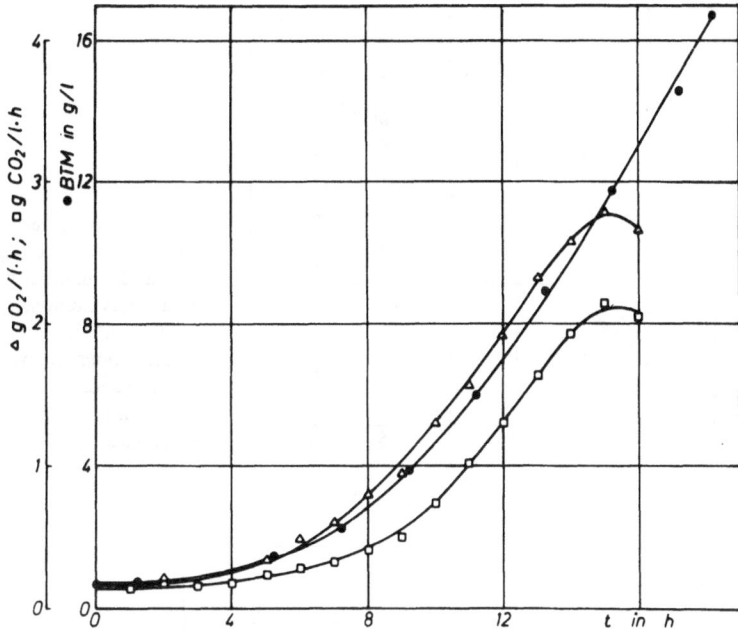

Fig. 26. Growth curve of *Candida boidinii* on 0.56 g · l⁻¹ ethanol. Porous plate
- ● (dry) biomass *BTM*   (g · 1⁻¹)
- △ oxygen consumption   (g · 1⁻¹ · h⁻¹)
- □ CO₂ production   (g · 1⁻¹ · h⁻¹)

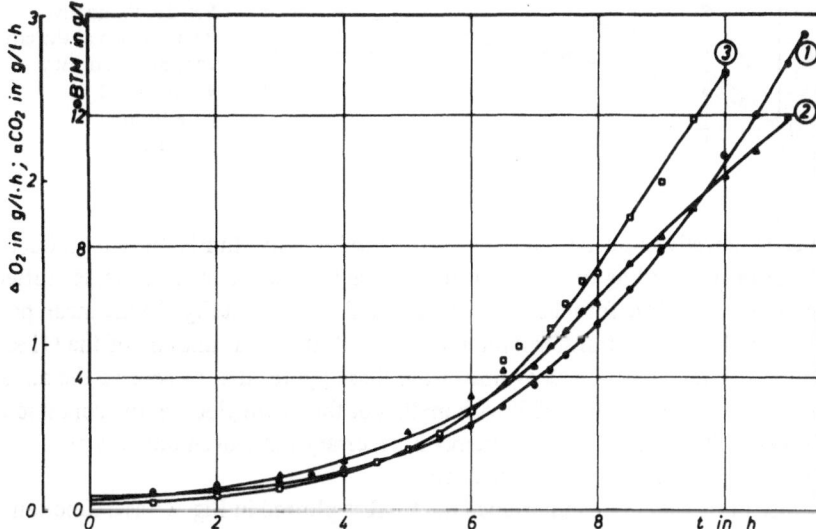

Fig. 27. Growth curve of *Candida boidinii* on glucose. Porous plate.
- *1* ● (dry) biomass (g · 1⁻¹)
- *2* ▲ oxygen consumption (g · 1⁻¹ · h⁻¹)
- *3* □ CO₂ production   (g · 1⁻¹ · h⁻¹)

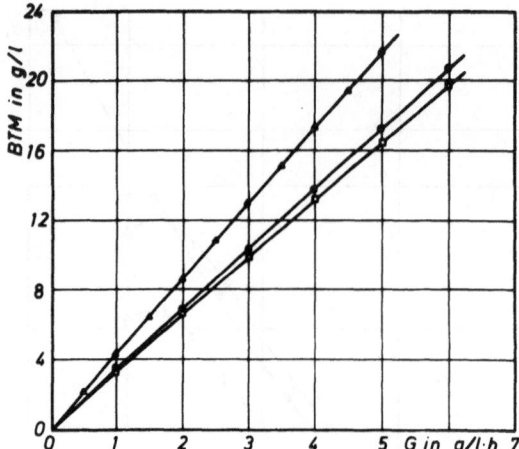

Fig. 28. Critical cell concentration at the beginning of oxygen transfer limitation (dry biomass $g \cdot l^{-1}$) as function of the maximum oxygen transfer rate $(g \cdot l^{-1} \cdot h^{-1})$.
Candida boidinii with
● methanol
□ ethanol
▲ glucose

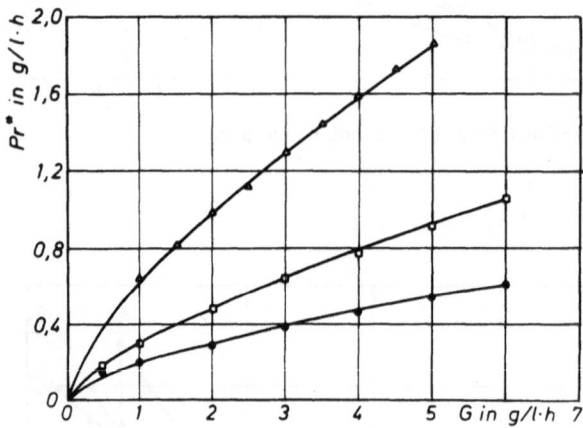

Fig. 29. Productivity Pr^* according to Gaden at the critical cell concentration with
● methanol
□ ethanol
△ glucose

An economically important quantity is the productivity. In Fig. 29 the productivity Pr^* (defined according to Gaden) at the critical biomass concentration, is plotted as a function of the *OTR* for methanol, ethanol and glucose. In Fig. 30 the mean productivity Pr in the oxygen transfer limited range, is plotted as a function of the OTR for the same substrates. It can be recognized from these figures that to achieve the same productivity requires a higher *OTR* with methanol than with glucose. In economic terms this means that a higher specific energy is necessary for a given biomass production when using methanol as compared with glucose.

Because of its economic importance the OTR is plotted in Fig. 31 as a function of time for media based on methanol, ethanol and glucose with porous and perforated plates. In spite of the high aeration in system with alcohol solutes and a perforated plate the OTR does not surpass 1.0. With a porous plate the maximum *OTR*'s amount to 2.5 for glucose, 3 for methanol and 5 for ethanol.

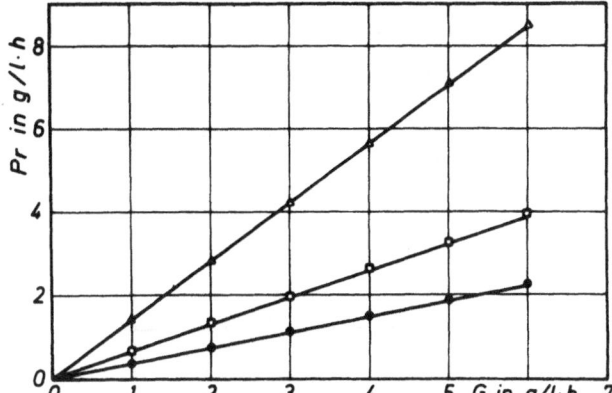

Fig. 30. Productivity Pr in the oxygen transfer limited range with

- ● methanol
- ▫ ethanol
- △ glucose

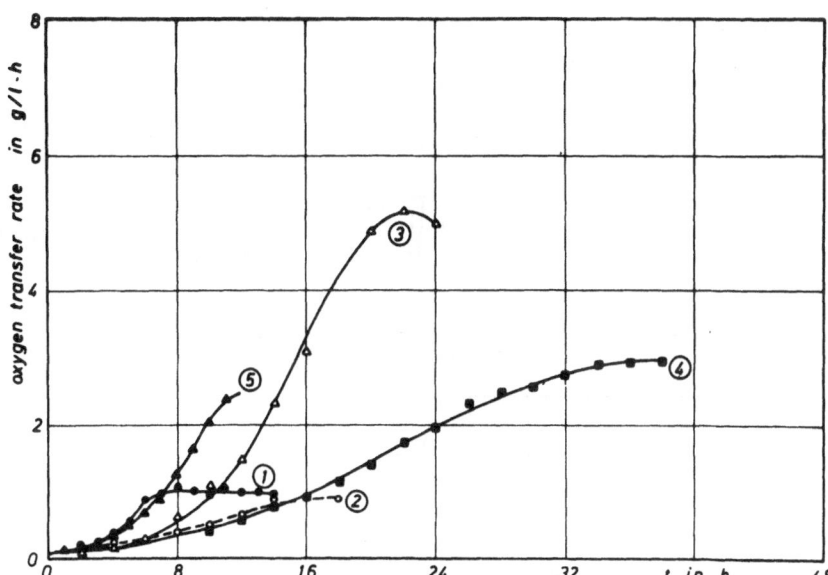

Fig. 31. Oxygen Transfer rate as function of the cultivation time.

		w_{SG} (cm · s^{-1})
1	ethanol-perforated plate*	3.7
2	methanol-perforated plate	3.7
3	ethanol-porous plate	1.7
4	methanol-porous plate	4.0
5	glucose-porous plate	2.9

* inoculum culture with ethanol

It is very instructive to compare the longitudinal concentration profiles of oxygen in the media at various biomass concentrations for methanol (Fig. 11), ethanol (Fig. 32), and glucose (Fig. 33) when using a porous plate aerator. In spite of the low maximum specific growth rate μ_m of the yeast cells on methanol the fall in concentration of the dissolved oxygen is fairly large due to the low oxygen yield coefficient and OTR in this medium (Fig. 11). Though μ_m for the yeast on ethanol is much higher than on methanol the fall in concentration of dissolved oxygen along the column is only slightly less than

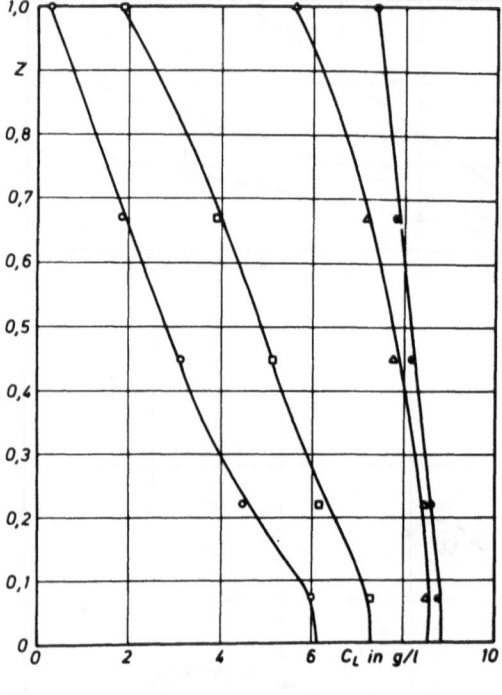

Fig. 32. Longitudinal concentration profiles of dissolved oxygen. *Candida boidinii* with ethanol and porous plate.
$w_{SG} = 1.7$ cm, $w_{SL} = 1.7$ cm $\cdot$ s^{-1}

time (h)	(dry) biomass (g $\cdot$ l^{-1})
● 2	0.85
△ 8	2.88
□ 12	6.52
○ 14	9.81

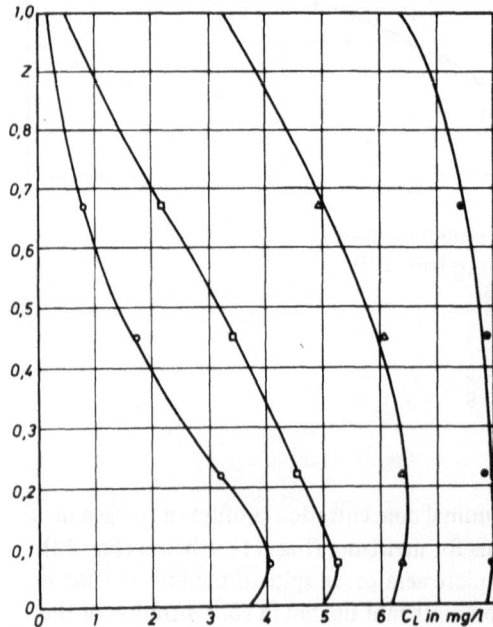

Fig. 33. Longitudinal concentration profiles of oxygen solute. *Candida boidinii* with glucose and a porous plate.
$w_{SG} = 2.86 - 3.57$ cm $\cdot$ s^{-1}, w_{SL} $= 1.74$ cm $\cdot$ s^{-1}

time (h)	(dry) biomass (g $\cdot$ l^{-1})
● 4	1.25
△ 6	2.55
□ 8	5.60
○ 10	10.75

with methanol because of the high oxygen yield coefficient and *OTR* which occurs in the ethanol based medium (Fig. 32).

In glucose based media a high *OTR* can be achieved and the yeast has a very high oxygen yield coefficient; nevertheless the fall in oxygen concentration is the largest in comparison with the methanol and ethanol based media at the same biomass concentration (Fig. 33). This can be attributed to the very high μ_m of the yeast on glucose. Figure 34

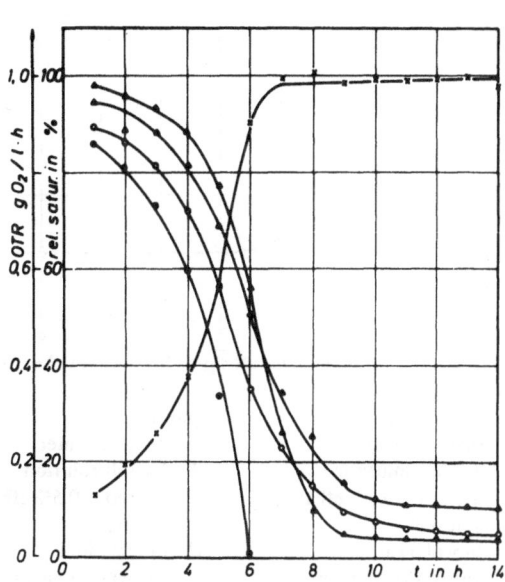

Fig. 34. Relative oxygen saturation (%) and oxygen transfer rate $(g \cdot l^{-1} \cdot h^{-1})$ as a function of the fermentation time. *Candida boidinii* on ethanol. Perforated plate, $w_{SG} = 2.98{-}3.57$ cm $\cdot$ s^{-1}, $w_{SL} = 1.74$ cm $\cdot$ s^{-1}

Relative position of the sampling (aerator plate at $z = 0$)

● $z = -\ 0.04$
○   0.07
▲   0.22
△   0.45
✕   oxygen transfer rate

clearly shows how, for ethanol based media the relative oxygen saturation falls and the *OTR* increases with increasing fermentation time. At about 7 h the *OTR* reaches a constant value completely corresponding to oxygen transfer limited growth.

To elucidate the parameters which were responsible for the differences in the *OTR*'s in the various media, the bubble size distributions were measured during growth. In Figs. 16, 35, 36 the mean bubble diameters d_B are plotted as a function of the biomass concentration for methanol and perforated and porous plates (Fig. 16), for ethanol and the same plates (Fig. 35), and for glucose and a porous plate (Fig. 36). As in the case of methanol (Fig. 16) the type of gas distributor strongly influences d_B for ethanol based media. The mean bubble diameters are much smaller with a porous plate (2) than with a perforated plate (1) (Fig. 35). The mean bubble diameter increases with a porous plate after the inoculation, passes through a maximum and approaches a constant value, which is slightly higher than the mean bubble diameter in the absence of cells (c).

A maximum in the d_B after the inoculation also occurs when glucose media is used (Fig. 36). The slight influence of the substrate on the mean diameter with a perforated plate is shown in Fig. 37. A more significant substrate effect was found with a porous plate (Fig. 38), where a smaller d_B was obtained with ethanol (1) than with methanol (2) in the presence of cells. This effect is similar to that occurring with cell free media (c) and (e).

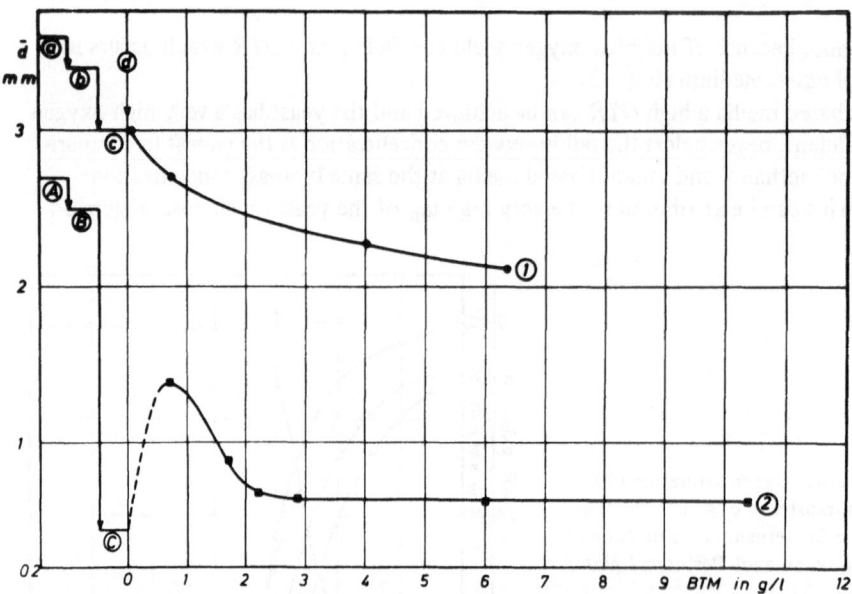

Fig. 35. Mean bubble diameter $\overline{d}_B$ as function of the (dry) biomass concentration (g · l⁻¹). *Candida boidinii* on ethanol.

perforated plate		porous plate	
a	distilled water	A	distilled water
b	1% salt solution	B	1% salt solution
c	1% salt + 0.5% ethanol solution	C	1% salt + 0.5% ethanol solution
d	inoculation	d	inoculation
1	cultivation medium	2	cultivation medium.

a–c: $w_{SG} = 4.5$ cm · s⁻¹   A–C: $w_{SG} = 2$ cm · s⁻¹

1:     $w_{SG} = 4.5–6.0$ cm · s⁻¹   2: $w_{SG} = 2.3–4.2$ cm · s⁻¹

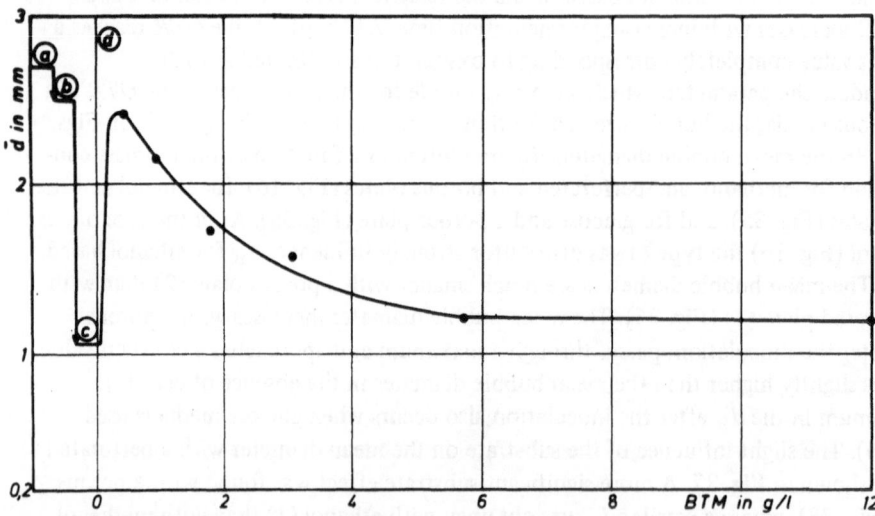

Fig. 36. Mean bubble diameter d_B as function of the (dry) biomass concentration (g · l⁻¹). *Candida boidinii* on glucose. Porous plate.

a	distilled water	c	1% salt + 0.5% glucose solution
b	1% salt solution	d	inoculation

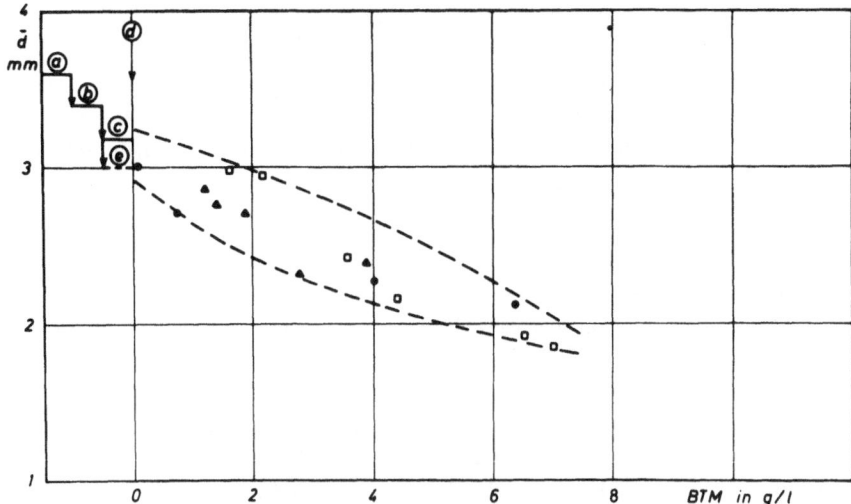

Fig. 37. Comparison of the mean bubble diameter d_B in methanol and ethanol media as a function of the dry biomass concentration (g/l). *Candida boidinii*, perforated plate.

a   distilled water
b   1% salt solution
c   1% salt + 0.5% methanol solution
d   inoculation

e   1% salt + 0.5% ethanol solution
•   ethanol substrate w_{SG} = 4.5–6.0 cm · s^{-1}
▲   ethanol substrate w_{SG} = 3.6–3.9 cm · s^{-1}
□   methanol substrate w_{SG} = 4.1–6.4 cm · s^{-1}

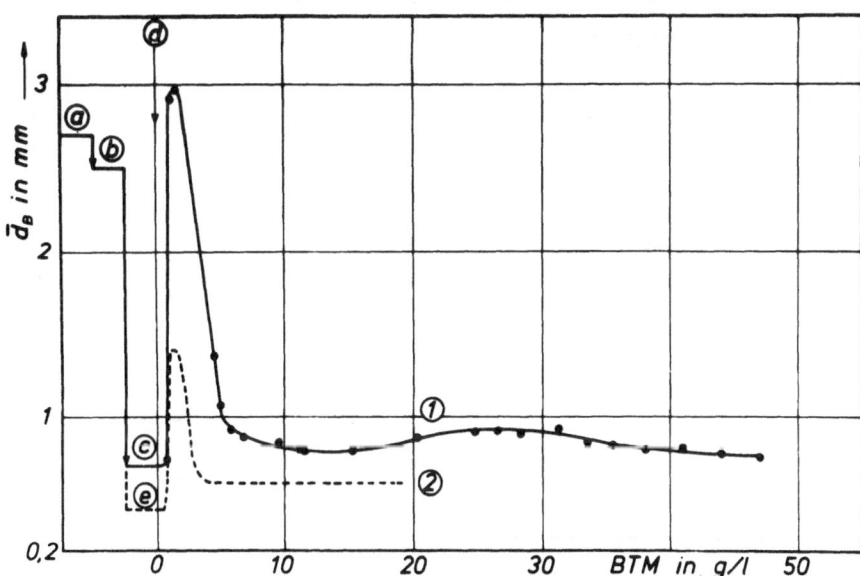

Fig. 38. Comparison of the mean bubble diameter $\overline{d}_B$ in methanol and ethanol media as a function of the dry biomass concentration (g · l^{-1}). *Candida boidinii*, porous plate.

a   distilled water
b   1% salt solution
c   1% salt + 0.4% methanol solution
d   inoculation

e   1% salt + 0.4% ethanol solution
1   growth medium with methanol
2   growth medium with ethanol

Table 11 shows once again that no "bubbling gas" state occurs, but that a gas jet is formed and hence the Rayleigh relation can be applied. Since according to this relation d_{Bin} does not depend on the composition of the medium, the initial bubble sizes must be the same in all systems with the same aerator independent of substrate, providing no turbulence exists. The differences in $\bar{d}_B$ are then caused only by the different d_{Bmax} and the tendency for coalescence.

Table 11. Comparison of calculated initial bubble diameters with measured mean bubble diameters in growth systems using ethanol. $C_s = 5$ g · l^{-1}

Aerator type	d_{Bin} (cm) calculated according to		$\bar{d}_B$ (cm) measured at biomass concentration
	Kumar [13]	Rayleigh [65]	6 g · l^{-1}
Perforated plate	0.31	0.107	0.20
Porous plate	0.17	0.032	0.062

A comparison of the $\bar{d}_B$'s with methanol, ethanol and glucose is given in Fig. 39. In cell free systems the mean bubble diameters follow the sequence: $\bar{d}_B$ (ethanol) $< \bar{d}_B$ (methanol) $< \bar{d}_B$ (glucose). In the presence of cells at the higher cell concentrations the same sequence prevails.

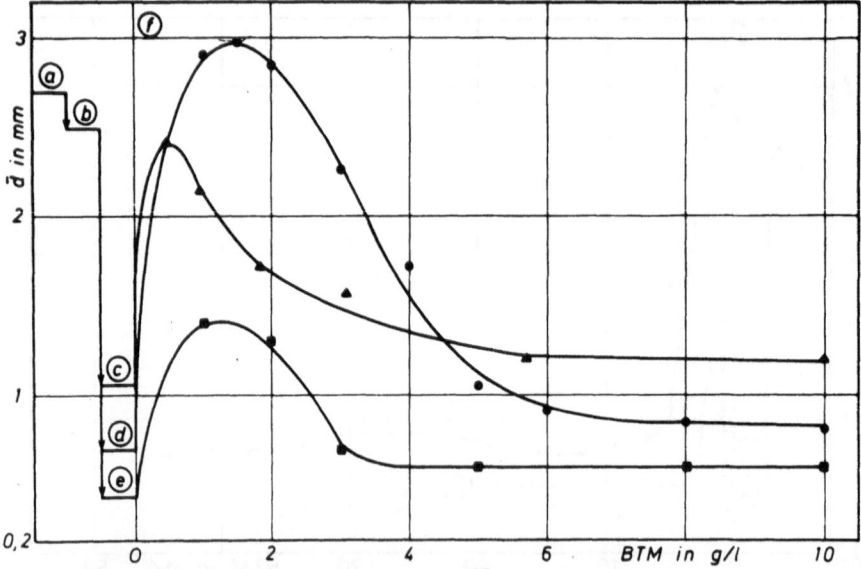

Fig. 39. Comparison of the mean bubble diameters $\bar{d}_B$ in methanol, ethanol and glucose medium as a function of the dry biomass concentration (g · l^{-1}). *Candida boidinii*, porous plate.

a   distilled water
b   1% salt solution
c   1% salt + 0.5% glucose solution
d   1% salt + 0.5% methanol solution
e   1% salt + 0.5% ethanol solution

f   inoculation
●   cultivation medium with methanol
▲   cultivation medium with glucose
■   cultivation medium with ethanol

The differences in $\bar{d}_B$ in cell free systems are due to the type of aerator and solute as discussed in detail in [1]. The influence of the aerator on $\bar{d}_B$ in presence of cells was considered in Sect. 6.1.

The initial bubble size at the aerator is smaller than the dynamical equilibrium bubble size in the column. Therefore the bubble size increases along the column and approaches the dynamical equilibrium bubble size, if the coalescence rate is high enough. Alcohols decrease, and salt and glucose increase the surface tension of water and hence they respectively decrease and increase the dynamical equilibrium bubble size. Furthermore, these solutes diminish the coalescence rate considerably. Since the coalescence suppressing effect of ethanol and glucose is higher than that of methanol the differences in the mean bubble diameter are increased by this effect, if methanol is compared with ethanol and they are diminished, if methanol and glucose are compared.

In the presence of cells the surface tension is much lower than in cell free systems. Therefore it is to be expected that the dynamical equilibrium bubble sizes are smaller than in the corresponding model media. Nevertheless the measured bubble diameters are larger, since the dynamical equilibrium bubble sizes are larger due to some damping of the turbulence near the surface covered by surfactants and proteins. Moreover, the coalescence rate is accelerated in the presence of the cells, apparently by the same extent for all three substrates, since the differences in $\bar{d}_B$ of the cell free systems for ethanol, methanol and glucose are preserved (Fig. 39).

By means of Eq. (8) the specific interfacial area A was calculated from the measured d_s and ϵ_G values. A comparison between Figs. 17 and 40 indicates that in bubble columns with methanol and ethanol solutes using a perforated plate the same specific interfacial areas are produced. During the fermentation a'/V increases due to the increasing cell concentration, and it approaches a constant value at high biomass density. The aeration rate also influences a'/V significantly (Fig. 40). In a bubble column reactor using metha-

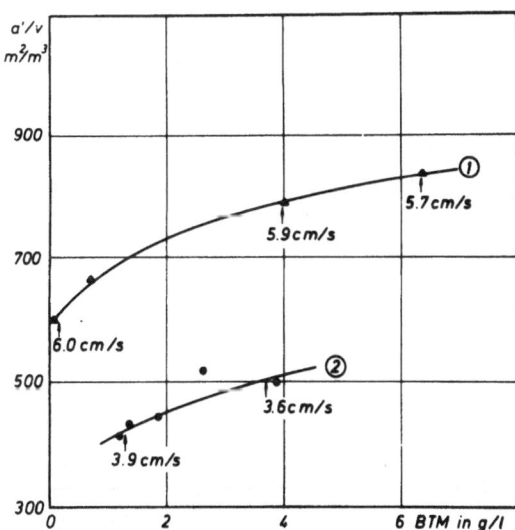

Fig. 40. Specific interfacial area a'/V as function of the dry biomass concentration. *Candida boidinii* on 0.5% ethanol with a perforated plate.
1 w_{SG} = 5.7–6.0 cm · s^{-1}
2 w_{SG} = 3.6–3.9 cm · s^{-1}
arrows indicate the change of the superficial gas velocity

nol at a high biomass concentration, a specific interfacial area of 1100 m^{-1} is achieved
by a perforated plate at $w_{SG} = 6.4 \text{ cm} \cdot \text{s}^{-1}$ (1.5 vvm).

In a bubble column with a porous plate aerator the substrate has a strong influence on
a'/V (Figs. 18, 41, and 42) which increase in the following sequence: glucose (1030 m^{-1}),
methanol ($1300–1500 \text{ m}^{-1}$), and ethanol (1700 m^{-1}) at high biomass concentrations
and rather low aeration rates (0.8 vvm for ethanol and glucose and 1.1 vvm for methanol).

To estimate the volumetric mass transfer coefficient ($k_L a$) the longitudinal concentration profiles for dissolved oxygen, calculated by means of the model described in
Sect. 4.2, were fitted to the measured data (Figs. 19, 20, 43, and 44). As pointed out
in Sect. 6.1, the calculated profiles were fitted to the measured data using a sole St-num-
ber for a perforated plate aerator and methanol (Fig. 19).

With a porous plate aerator and methanol the specific interfacial area decreases along
the column due to coalescence. Therefore a satisfactory curve-fit is only possible using
a position dependent St-number (Fig. 20). The same is true for glucose (Fig. 44). With
ethanol, which suppresses coalescence very effectively, less variation in the St-number
exists (Fig. 43).

A comparison between volumetric mass transfer coefficients (Table 12) indicates, as
expected from the investigations with model media [1], that the substrate significantly
influences ($k_L a$).

The highest ($k_L a$) values were obtained with ethanol based media with a porous plate
aerator.

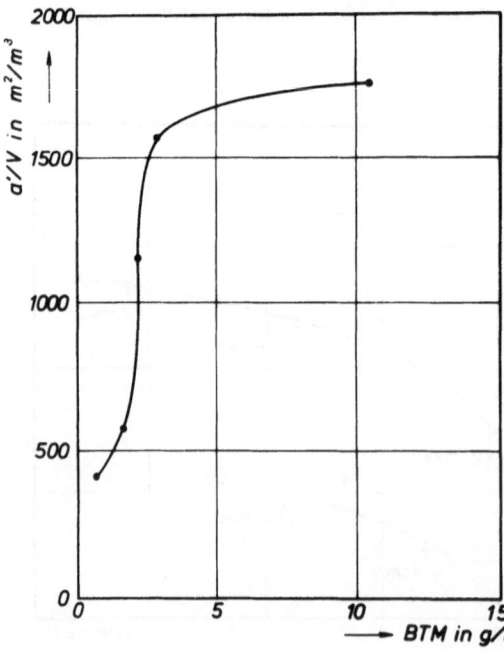

Fig. 41. Specific interfacial area a'/V
as function of the dry biomass concentration. *Candida boidinii* on 0.5%
ethanol with a porous plate.
$w_{SG} = 2.3 \text{ cm} \cdot \text{s}^{-1}$ (0.5 vvm)

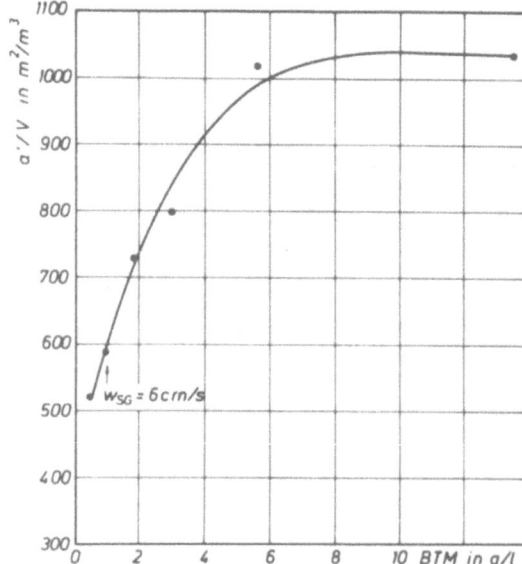

Fig. 42. Specific interfacial area a'/V as function of the (dry) biomass concentration. *Candida boidinii* on 0.5% glucose with a porous plate: $w_{SG} = 2.9–3.6$ cm $\cdot$ s^{-1} (with exception of the time shortly after the inoculation (see arrow))

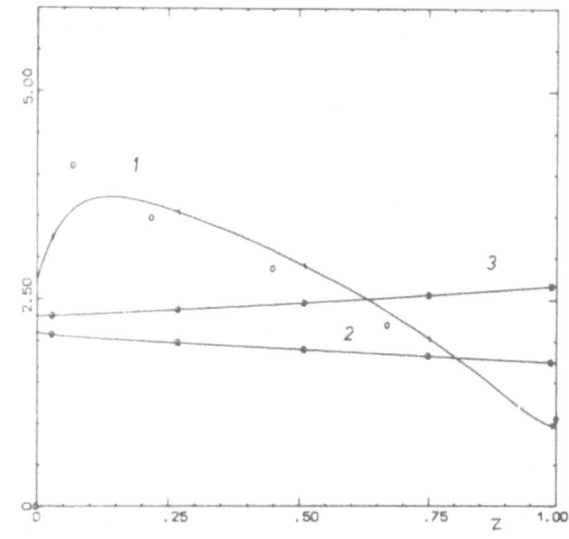

Fig. 43. Fitted longitudinal concentration profiles for dissolved oxygen. *Candida boidinii* on ethanol. Porous plate.
$w_{SG} = 1.7$ cm $\cdot$ s^{-1},
$w_{SL} = 1.7$ cm $\cdot$ s^{-1}
1 mole fraction of oxygen in the medium $x_{Lo} \cdot 10^6$
2 mole fraction of oxygen in the gas $x_{LG} \cdot 10$
3 superficial gas velocity (cm $\cdot$ s^{-1})

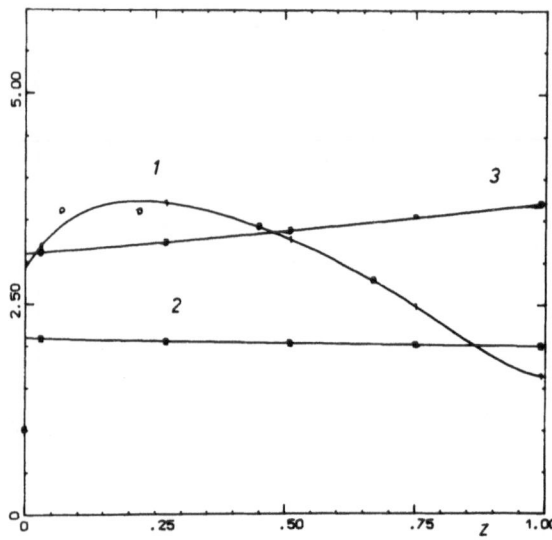

Fig. 44. Fitted longitudinal concentration profiles for dissolved
oxygen. *Candida boidinii* on glucose. Porous plate,
$w_{SG} = 3.3 \cdot cm \cdot s^{-1}$ ·
$w_{SL} = 1.7 \, cm \cdot s^{-1}$
1   mole fraction of oxygen
      in the medium $x_{Lo} \cdot 10^6$
2   mole fraction of oxygen
      in the gas $x_{LG} \cdot 10$
3   superficial gas velocity
      $w_{SG} \, (cm \cdot s^{-1})$

Table 12. Comparison of the volumetric mass transfer coefficients ($k_L a$) and mass transfer coefficients k_L in various antifoam free growth media using perforated plate PE and porous plate PO
aerators. $C_s = 5 \, g \cdot l^{-1}$. Mechanical foam separator

	Substrate			
	Methanol		Ethanol	Glucose
Aerator	PO	PE	PO	PO
w_{SG} (cm s^{-1})	3.8	6.3	2.3	3.3
$k_L a$ (s^{-1})	0.138[a]	0.0818	0.320[a]	0.130
$k_L \cdot 10^2$	0.72[b]		1.9[b]	0.98[b]
(cm · s^{-1})	0.56[c]	0.59	1.2[c]	0.56[c]

[a] Valid only near to the aerator.   [c] Calculated by $\dfrac{(\overline{k_L a})}{\overline{a}}$.

[b] Calculated by $\dfrac{(k_L a)}{\overline{a}}$.

The mass transfer coefficients k_L in Table 12 were calculated from the initial ($k_L a$) and
the mean specific interfacial area a or from the mean ($\overline{k_L a}$) and a for a porous plate
aerator. Since $k_L a$ varies in systems with a porous plate aerator mainly due to the reduction of the interfacial area with distance, k_L should remain nearly constant. The true
k_L values are between these limits and probably nearer to the lower limit. A comparison
of k_L values with those measured using the corresponding model media [1] indicates
that the latter are higher by a factor of 5–10. This significant change of k_L is caused
by the alteration of the properties of the gas/liquid interfacial area due to surface active
substances and proteins. The latter influences ($k_L a$) mainly due to fluid dynamic effects.

A comparison of the k_L values in Table 12 with those measured by Koide *et al.* [102] in water systems with surfactants shows excellent agreement. Furthermore, a comparison of the k_L values in growth media with those measured in the model media in the Levenspiel-stirred-cell (Table 1 in [1]) indicates that the latter are lower ($0.2–0.4 \cdot 10^{-2}$ cm s^{-1}) probably because of the more marked retardation of surface flow and damping of turbulence at the interface.

The higher k_L with ethanol rather than with methanol and glucose can be explained by the higher surface activity of the former. The surface active substances, which retard the surface flow, are more effectively displaced by ethanol than by methanol and glucose.

Table 13. Comparison of the biomass productivities Pr in the oxygen transfer limited range in various antifoam free growth media with porous and perforated plate aerators. Mechanical foam separator

	Substrate				
	Methanol		Ethanol		Glucose
Aerator	PE	PO	PE	PO	PO
Pr	0.31	0.68	0.62	1.63	2.86
w_{SG} (cm s^{-1})	3.81	3.81	3.20	2.30	3.30
$\frac{E}{V_L}$ $\left(\frac{kW}{m^3}\right)$	0.43	0.43	0.37	0.26	0.37
$\frac{E^*}{V_L}$ $\left(\frac{kW}{m^3}\right)$	< 1[a]	< 1[a]	< 5[a]	< 5[a]	< 1[a]

[a] Depends on the biomass concentration.

The productivities Pr, which were measured with various substrates and aerators, are compiled in Table 13.

The particular substrate and aerator strongly influence Pr. At fairly low specific energy requirements high productivities were achieved, especially with glucose and ethanol in columns with a porous plate aerator.

The growth media, which have been considered up to this point, consisted of chemically well-defined (so called "synthetic") ingredients. However, in the fermentation industry "complex" ingredients of ill-defined composition are used. In view of this situation the influence of corn steep liquor as an example of a common complex ingredient of low viscosity was considered. A medium which is used to produce glucose oxidase from *Aspergillus niger* consists of

Dextrose	$65.70 \text{ g} \cdot l^{-1}$
CaCO$_3$ (powder)	$12.80 \text{ g} \cdot l^{-1}$
NH$_4$NO$_3$	$0.63 \text{ g} \cdot l^{-1}$
Mg^{2+} and K$^+$ sources	$0.56 \text{ g} \cdot l^{-1}$
Corn steep liquor	$13.00 \text{ g} \cdot l^{-1}$.

The composition of corn steep liquors has been investigated by Cejka [98].

The same perforated plate and injector nozzle were applied as for yeast cell mass production, though only the mean relative gas hold-up ϵ_G, which is characteristic of the Sauter mean bubble diameter [1], was measured.

In Figs. 45–47 ϵ_G is plotted as a function of the superficial gas velocity for a perforated plate and a mineral salts solution (Fig. 45) with corn steep liquor (Fig. 46) and with different amounts of corn cob oil (Fig. 47); data on tap water are included for comparison.

Salt additives raise ϵ_G because of the hindrance of coalescence [1]. Corn steep liquor causes a strong increase of ϵ_G especially at high gas velocities both due to the reduction of the surface tension and consequently the dynamical equilibrium bubble diameter, and the coalescence depression mainly due to the acid (lactic acid and aminoacid) content. Foam formation was observed at $w_{SG} = 8$ cm $\cdot$ s^{-1} as a result of the surface active substances and proteins present in corn steep liquor. The addition of a small amount of oil reduces ϵ_G dramatically (Fig. 47). The oil forms a monomolecular film at the gas/liquid interface, promotes coalescence and increases the dynamical equilibrium bubble size as discussed in section 6.3. With increasing gas flow rate slugging occurs at $w_{SG} = 6$ cm $\cdot$ s^{-1}. In Fig. 48 the influence of the ingredients on ϵ_G are compared for $w_{SL} = 2.2$ cm $\cdot$ s^{-1}. With addition of salt and corn steep liquor ϵ_G increases and with oil it decreases. Even at low gas velocities the complete culture media (with oil) yields higher ϵ_G than tap water.

In Fig. 49 the ϵ_G values are compared for the injector nozzle with water (1) salt and corn steep liquor (2), a small amount of oil (3) and 30 ml oil (4).

It can be recognized that at low gas flow rates the influence of oil addition is small and high ϵ_G values can be achieved. However, with oil, slugging occurs at $w_{SG} = 4$ cm/s and without oil at $w_{SG} = 5.5$ cm $\cdot$ s^{-1}.

The systems without oil and corn steep liquor are compared in Fig. 50. This data indicates the much better performance of the injector nozzle than the perforated plate, especially at low liquid velocities.

With corn steep liquor and oil addition the injector nozzle still yields higher ϵ_G than the perforated plate, especially at $w_{SG} = 3$ cm $\cdot$ s^{-1} and at low liquid velocities (Fig. 51).

A systematic investigation of systems containing complex media would certainly enable us to find much better operating conditions.

This example shows that fairly high performance of bubble column fermentors can be achieved with complex growth media if the influence of the process parameters is established.

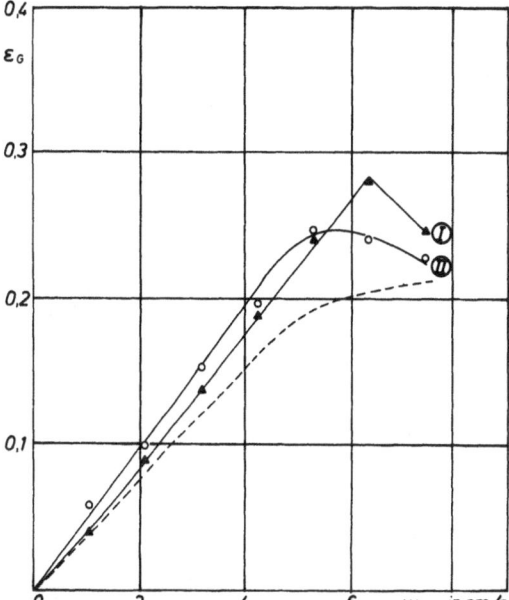

Fig. 45. Relative mean gas hold up as
a function of the superficial gas velocity.
Perforated plate.
--- H_2O
▲ (I) salt solution
       $w_{SL} = 1.2$ cm · s^{-1}
o (II) salt solution
       $w_{SL} = 2.2$ cm · s^{-1}

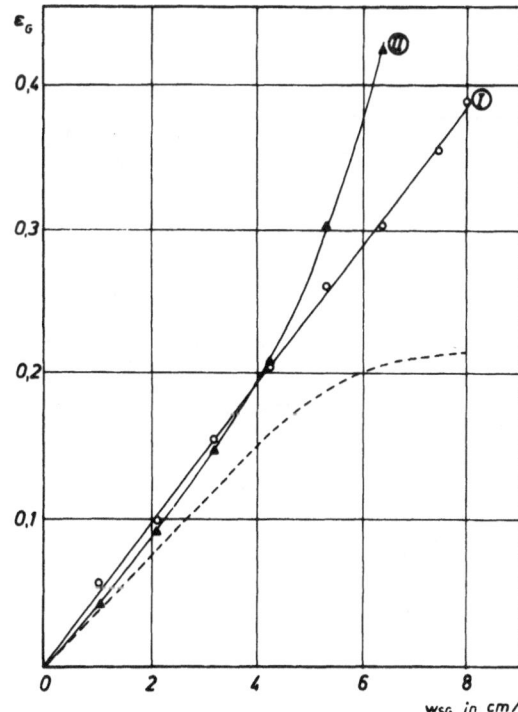

Fig. 46. Relative mean gas hold up as
a function of the superficial gas
velocity. Perforated plate.
--- H_2O
o (I) salt + corn steep liquor,
       $w_{SL} = 2.2$ cm · s^{-1}
▲ (II) salt + corn steep liquor
       $w_{SL} = 0.8$ cm · s^{-1}

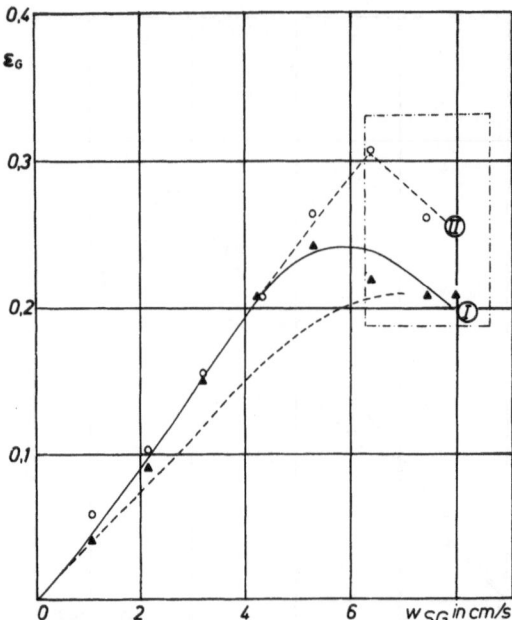

Fig. 47. Relative mean gas hold up as a function of the superficial gas velocity. Perforated plate.

--- H_2O

(I) salt + corn steep liquor + corn cob oil (8 droplets), $w_{SL} = 0.8$ cm $\cdot$ s^{-1}

(II) salt + corn steep liquor + corn cob oil (8 droplets), $w_{SL} = 2.2$ cm $\cdot$ s^{-1}

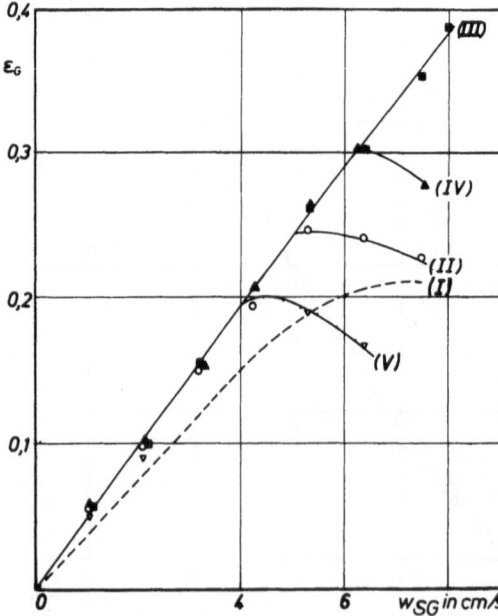

Fig. 48. Comparison of the influence of various ingredients on the mean relative gas hold up. Perforated plate. $w_{SL} = 2.2$ cm $\cdot$ s^{-1}

--- (I) H_2O

o   (II)   salt solution

■   (III)   salt solution + corn steep liquor

▲   (IV)   salt + corn steep liquor + corn cob oil (8 drplets)

▽   (V)   salt + corn steep liquor + cob oil (30 ml)

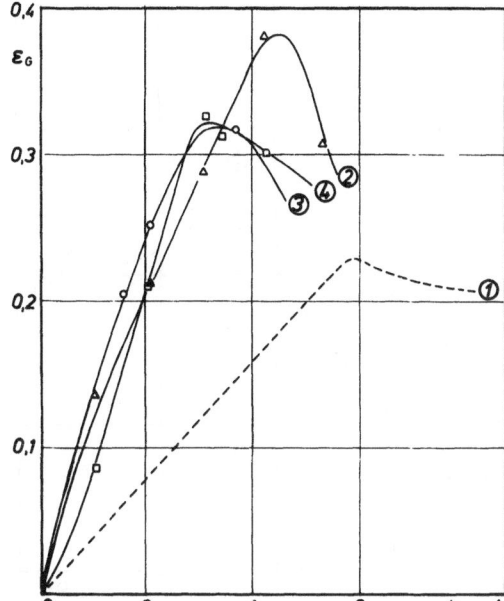

Fig. 49. Comparison of the influence of various ingredients on the mean relative gas hold up. Injector nozzle, $w_{SL} = 2.2$ cm · s^{-1}

---   *1* H_2O
△   *2*   salt solution
○   *3*   salt + corn steep liquor
           + corn cob oil (8 droplets)
□   *4*   salt + corn steep liquor
           + corn cob oil (30 ml)

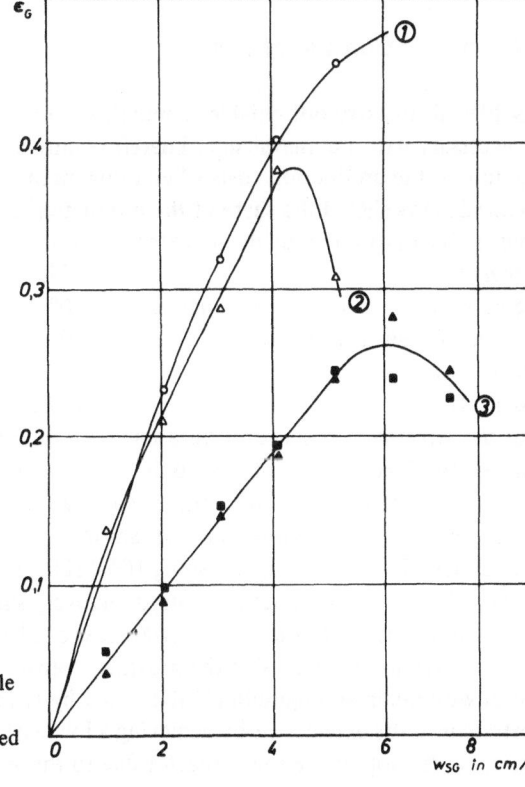

Fig. 50. Comparison of the influence of various aerators on the relative mean gas hold up. Salt solution.

○   *1* $w_{SL} = 1.2$ cm · s^{-1} ⎫
△   *2* $w_{SL} = 2.2$ cm · s^{-1} ⎬ injector nozzle
                               ⎭
■ $w_{SL} = 2.2$ cm · s^{-1} ⎫
▲ $w_{SL} = 1.2$ cm · s^{-1} ⎬ (3) perforated
                                 plate

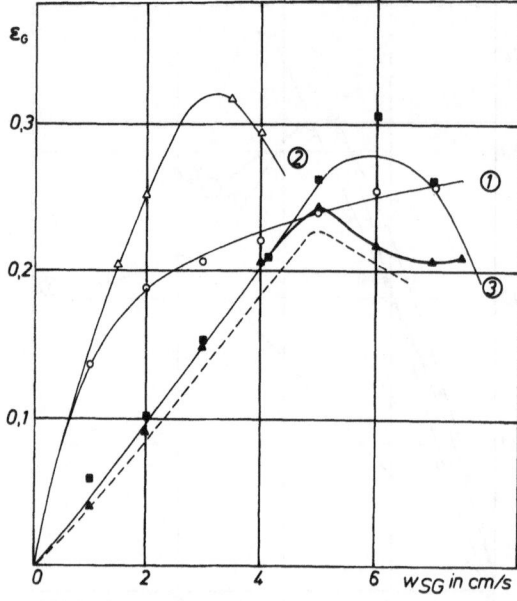

Fig. 51. Comparison of the influence of various aerators on the relative mean gas hold up. Salt + corn liquor + corn oil (8 droplets).

o *1* $w_{SL} = 0.8$ cm · s⁻¹ } injector
△ *2* $w_{SL} = 2.2$ cm · s⁻¹ } nozzle
▲ $w_{SL} = 0.8$ cm · s⁻¹ } (3) perforated
■ $w_{SL} = 2.2$ cm · s⁻¹ } (3) plate
– – – H_2O injector nozzle

7. Economic Considerations

It is difficult to carry out reliable economic calculations on cell mass production, because the necessary data are unavailable. Therefore only a rough estimate will be made of the structure of the production costs when using methanol. According to the European Chemical News (ECN) the share of the production costs of SCP in a 100000 t/a plant using methanol can be roughly allocated as follows [99]:

Methanol:	52%
Energy + water + aeration + auxiliaries	10%
Personnel + maintenance	9%
Capital costs	19%
Chemicals	10%.

Thus the investment costs contribute some 20% and the energy some 10% of the overall costs. Of the 20% investment costs only a share, approximately 20 to 25% is due to the reactor system (reactor 5–10%, stirrer and aerator 5–10%, cooling 10–15%); a higher proportion of 25–40% is due to the separation and finishing stages (separation 10–15%, pretreatment 5–15%, drying, finishing 10% [100]. Because the methanol participates by about 50% of the production costs, a relatively small reduction in substrate loss can have a significant influence. The substrate loss can be diminished compared with conventional systems by increasing the substrate conversion by using a continuous tower reactor with minimal longitudinal mixing and by reducing the evaporative loss of the substrate into the exhaust air by achieving a low exit concentration of substrate in the medium at the top of the tower reactor due to this higher conversion. The substrate loss can be further reduced by taking advantage of the low aeration rates characteristics of tower reactors. In the present article the optimization of the substrate conversion was not discussed, since not enough experimental data are available. However, it should be pointed out that the application of a continuous reactor with low intensity of longitudinal mixing, like a high bubble column with methanol growth medium [1] yields much higher substrate conversions than a continuous stirred tank reactor with the same volume and dilution rate [107]. It can also be shown that the substrate conversion can be maximized by means of an optimum recycling ratio of the substrate free and cell containing medium [107].

Since the reactor investment costs contribute about 4–5% and the reactor operating costs contribute about 10%, the type of the bioreactor can only influence 15% of the total production costs due to the reduction of the investment and operation costs. Since the reduction in the investment costs by use of a tower reactor is relatively slight, the main advantage of tower reactors appears in the lower operating costs, especially in the reduced energy and cooling water costs. Because of the small difference between the required growth temperature and the temperature of the cooling water (especially in summer), the specific cooling water requirement is very high. Thus the reduction of the specific energy intake reduces the running costs by diminishing the cooling water requirement. In large plants the availability of a large amount of cooling water could limit the production capacity especially in the summer and it could be necessary to cool the cooling water. Thus the application of tower reactors provides several possibilities for the reduction in the operation costs compared with conventional systems.

The influence of the tower reactor on product quality has not been considered, though this topic undoubtedly is important for successful large scale operation.

8. Outlook

The results presented indicate that tower reactors are economically attractive for cell mass production using alcohols. The results have been obtained with low viscosity synthetic media and cannot be extended to complex media because of the influence of the physical properties and ingredients of such media, e.g., viscosity effects of natural ingredients.

Because antifoam agents mostly cancel the attractive properties of tower reactors, it is advisable to avoid using such additives, if one wants to benefit from their economical advantages. To renounce antifoam agents completely, it is necessary to carry out systematic investigations upon foam formation and control with practical growth systems.

Acknowledgement

The authors gratefully acknowledge the financial support of the "Bundesministerium für Forschung und Technologie" of German Federal Government.

Nomenclature

(L = length-, M = mass-, T = time units)

A'	surface area (general)	L^2
$A = \dfrac{a'}{V}$	specific surface area gas/liquid	L^{-1}
$a = \dfrac{a'}{V_L}$	specific surface area gas/liquid	L^{-1}
a'	surface area gas/liquid	L^2
a^*	radius of the jet	L
B	constant	T^{-1}
$Bo = \dfrac{w_L H}{D_L}$	Bodenstein number	–
$C_L,\ C_L^*$	concentration of oxygen in the bulk phase, at the interface	$M\,L^{-3}$
C_s	substrate concentration	$M\,L^{-3}$
D_m	molecular diffusivity	$L^2\,T^{-1}$
$D_{L\,ax}$	longitudinal dispersion coefficient in the liquid	$L^2\,T^{-1}$
D_{LB}	back mixing coefficient in the liquid	$L^2\,T^{-1}$
d_B	bubble diameter	L
d_S	Sauter mean bubble diameter	L
E	rate of energy dissipation, power input	$M\,L^2\,T^{-3}$
E/V_L	specific energy dissipation rate or specific power input due to compression	$M\,L^{-1}\,T^{-3}$
$\dfrac{E^*}{V_L}$	specific power input due to mechanical foam separation	$M\,L^{-1}\,T^{-3}$
$G = \dfrac{Q}{V}$	oxygen intake, oxygen transfer rate, OTR	$M\,T^{-1}\,L^{-3}$
$G_{max} = K_L a\, C_L^*$		$M\,T^{-1}\,L^{-3}$
H	height of the bubbling layer	L
H_L	height of the bubble free layer	L
k	wave number of disturbance	L^{-1}
K_s	saturation constant in the Monod equation	$M\,L^{-3}$
k_L	mass transfer coefficient	$L\,T^{-1}$
M_s	mole mass of component j	M
P	pressure	$M\,L^{-1}\,T^{-2}$
$Pr^* = \dfrac{x_t - x_o}{t - t_o}$	productivity according to Gaden	$M\,L^{-3}\,T^{-1}$
$Pr = \dfrac{dx}{dt}$	productivity	$M\,L^{-3}\,T^{-1}$
Q	mass transfer rate	$M\,T^{-1}$
r	reaction rate	$M\,L^{-3}\,T^{-1}$
$St = k_L a\,\bar{t}$	Stanton number	–
T	temperature	
t	time	T
$\bar{t}$	mean residence time of the liquid phase	T
X	cell mass	M
x	longitudinal coordinate	L
x_{Lo}	mole fraction of oxygen in the liquid phase	–
x_{LG}	mole fraction of oxygen in the gas phase	
Y	yield coefficient	–
$V,\ V_L$	volume of bubbling layer, bubble free layer	L^3
w_R	relative velocity of the bubble swarm	$L\,T^{-1}$

w_{SG}	superficial gas velocity at 20 °C 2 bar	$L\ T^{-1}$
$w_{SG}{}^*$	superficial gas velocity at 0 °C 1 bar	$L\ T^{-1}$
w_{SL}	superficial liquid velocity	$L\ T^{-1}$
w_G	effective gas velocity	$L\ T^{-1}$
w_L	effective liquid velocity	$L\ T^{-1}$
$We = \dfrac{\tau d_B}{\sigma}$	Weber number	–
$Z = \dfrac{x}{L}$	dimensionless longitudinal coordinate	–
α	growth rate of the disturbance	T^{-1}
ϵ_G	mean relative gas hold up	–
$\epsilon_L = (1 - \epsilon_G)$	mean relative liquid hold up	–
κ	compression modulus	$M\ L^{-1}\ T^{-2}$
μ	specific growth rate	T^{-1}
μ_m	maximum specific growth rate	T^{-1}
η_L	dynamic viscosity	$M\ L^{-1}\ T^{-1}$
ν_L	kinematic viscosity	$L^2\ T^{-1}$
π	surface pressure	$M\ L^{-1}\ T^{-2}$
ρ_L	density of liquid	$M\ L^{-3}$
σ	surface tension	$M\ T^{-2}$
τ	turbulent shear stress	$M\ L^{-1}\ T^{-2}$
BTM	(dry) biomass concentration	$M\ L^{-3}$
PE	perforated plate aerator	
PO	porous plate aerator	

References

1. Schügerl, K., Oels, U., Lücke, J.: Bubble column bioreactors. In: Advances in Biochemical Engineering, Vol. 7, p. 1. Berlin, Heidelberg, New York: Springer 1977.
2. Padday, J. F.: The Theory of Surface Tension. In: Surface and Colloid Science. Vol. 1, p. 39: Matijevic, E. (Ed.) New York: Wiley-Intersicence 1969.
3. Joly, M.: Rheological properties of Monomolecular Films. In: Surface and Colloid Science, Vol. 1, p. 1. Matijevic, E. (Ed.) New York: Wiley-Intersicence, 1969.
4. Davies, J. T., Rideal, E. K.: Interfacial Phenomena. New York: Acad. Press 1963.
5. Davies, J. T.: Mass Transfer and Interfacial Phenomena. In: Advances in Chemical Engineering. Drew, T. B. et al. (Eds.) p. 1. New York: Acad. Press 1963.
6. De Vries, A. J.: Foam Stability, Rubber Sticking. Delft 1957.
7. Schwuger, M. J.: Chemiker Zeitung 96, 248 (1972).
8. De Vries, A. J.: Proc. 2nd. Internat. Congr. Surface Activity 1, 256. London: Butterworths 1957.
9. Cumper, C. W. N.: Trans. Faraday Soc. 49, 1360 (1953).
10. Lücke, J., Oels, Schügerl, K.: Chem. Ing. Techn. 48, 573 (1976).
11. Goldacre, R. J.: Surface films in Surface Phenomena in Chemistry and Biology. Damielli, J. F., et al. (Eds.), pp. 278. Oxford: Pergamon Press 1958.
12. Gifford, W. A., Scriven, L. E.: Chem. Eng. Sci. 26, 287 (1971).
13. Kumar, R., Kuloor, N. R.: Adv. Chem. Engng. Vol. 8. Drew, T. B. et al. (Eds.). New York: Acad. Press 1970.
14. Calderbank, P. H.: Mass transfer in Fermentation Equipments. In: Biochemical and Biological Engineering Science, Vol. 1, p. 102. N. Blakebrough (Ed.) New York: Acad. Press 1967.
15. Kolmogoroff, A. N.: Compt. rend. acad. sci. U. S. S. R. 30, 301 (1941); see also in "Turbulence" classic papers on statistical theory. Friedlander, S. K., Topper, L., (Eds.) p. 159. New York: Interscience 1961.

16. Batchelor, G. K.: Camb. Phil. Soc. **47**, 359 (1951).
17. Levich, V. G.: Physiochemical Hydrodynamics. Englewood Cliffs, N. J.: Prentice Hall 1962.
18. Davis, R. E., Acrivos, A.: Chem. Eng. Sci. **21**, 681 (1966).
19. Saville, D. A.: The Chemical Engineering Journal **5**, 251 (1973).
20. Ruckenstein, E.: Chem. Eng. Sci **19**, 505 (1964).
21. Blank, M.: J. Phys. Chem. **65**, 1698 (1961).
22. Davies, J. T., Mayers, G. R. A.: Chem. Eng. Sci. **16**, 55 (1961).
23. Baird, M. H. I., Davidson, J. H.: Chem. Eng. Sci. **17**, 87 (1962).
24. Garner, F. H., Skelland, A. H. P.: Chem. Eng. Sci. **4**, 149 (1955).
25. West, F. B., *et al.*: Ind. Eng. Chem. **43**, 234 (1951); **44**, 625 (1952).
26. Garner, F. H., Hale, A. R.: Chem. Eng. Sci. **2**, 157 (1953).
27. Hancher, C. W., Thacker, L. H., Phares, E. F.: Biotechn. Bioeng. **16**, 475 (1974).
28. Bucholz, R.: Diplomarbeit, Technical University, Hanover 1976.
29. Lücke, J.: Doctoral Thesis, Technical University, Hanover 1976.
30. Electrolux Environmental Systems Division.
31. Marucci, G., Nicodemo, H.: Chem. Eng. Sci. **22**, 1257 (1967).
32. Zieminski, S. A., Caron, M. M., Blackmore, R. B.: Ind. Eng. Chem. Fundamentals **6**, 233 (1967).
33. Jackson, R.: Chem. Eng. No. **178**, 107 (1964).
34. Addison, C. C.: J. Chem. Soc. 535 (1943); 252, 477 (1944); 98 (1945).
35. Brown, A. G., Thumann, W. C., and McBain, I. W.: J. Colloid. Sci. **8**, 491, 508 (1953).
36. Sawistowski, H., Goltz, G. E.: Trans. Instn. Chem. Eng. **41**, 174 (1963).
37. Sawistowski, H.: Interfacial Phenomena. In: Recent advances in liquid-liquid extraction. Hanson, C. (Ed.) p. 293. London: Pergamon Press 1971.
38. Handbook of Chemistry and Physics. Hodgman, Ch. Weast, R. C., Selby, S. M. (Eds.). Cleveland: The Chemical Rubber Publishing Co. 1960.
39. Danckwerts, P. V.: Gas-liquid reactions. New York: McGraw Hill 1970.
40. Oels, U., Lücke, J., Schügerl, K.: Chem. Ing. Techn. **49**, 59 (1977).
41. Reuß, M., Piehl, Wagner, F.: Fifth International Fermentation Symposium Berlin 1976, Dellweg, H. (Ed.) p. 25.
42. Vogelmann, H., Eppert, K., Wagner, F.: Fifth International Fermentation Symposium, Berlin 1976, Dellweg, H. (Ed.) p. 28.
43. Vogelmann, H., Reuss, M., Gnieser, J., Wagner, F.: 3. Symposium Technische Mikrobiologie Berlin 1973, Inst. für Gärungsgewerbe und Biotechnologie, Dellweg, H. (Ed.) p. 215.
44. Zlokarnik, M. (article in this volume).
45. Oels, U., Doctoral Thesis, Technical University, Hanover 1975.
46. Todt, J., Doctoral Thesis, Technical University Hanover 1974.
47. Todt, J., Lücke, J., Schügerl, K., Renken, A.: Chem. Eng. Sci. **32**, 369 (1977).
48. Oels, U., Schügerl, K., Todt, J.: Chem. Ing. Techn. **48**, 73 (1976).
49. König, B.: Diplomarbeit, Technical University Hanover 1977.
50. Nagel, O., Kürten, H., Sinn, R.: Chem. Ing. Techn. **42**, 474 (1970).
51. Nagel, O., Kürten, H., Hegner, B., Sinn, R.: Chem. Ing. Techn. **42**, 921 (1970).
52. Brauer, H.: Grundlagen der Einphasen- und Mehrphasenströmungen, Aarau-Frankfurt/M.: Sauerländer 1971.
53. Ruff, K.: Chem. Ing. Techn. **46**, 769 (1974).
54. Luttmann, R.: Diplomarbeit, Technical University Hanover 1976.
55. Pratt, H. R. C.: Ind. Chemist **31**, 63 (1955).
56. Adler, I.: Diplomarbeit, Technical University Berlin 1975.
57. Marucci, G.: Chem. Eng. Sci. **24**, 975 (1964).
58. Deckwer, W. D., Burckhart, R., Zoll, G.: Chem. Eng. Sci. **29**, 2177 (1974).
59. Reuß, M.: Doctoral Thesis, Technical University Berlin 1970.
60. Reuss, M., Lehmann, J.: Application of multiphase models for biological processes. Presented on the Seminar "Methods of chemical kinetics and its application". Austrian working group "Chemisches Apparatewesen und Verfahrenstechnik" Graz. 25/26. Sept. 1975.

61. Reuss, M.: Fifth Internat. Ferment. Symp. Berlin 1976, p. 89.
62. Deckwer, W. D., Zaidi, A., Adler, I.: Chem. Ing. Techn. 49, 507 (1977).
63. Sahm, H., Wagner, F.: Arch. Microbiol. 84, 29 (1972).
64. Sahm, H., Wagner, F.: Euro. J. Biochem. 36, 250 (1973).
65. Tani, Y., Miya, T., Ogata, K.: Agr. Biol. Chem. 36, 76 (1972).
66. van Dijken, J. P.: Ph. D. Thesis, University Groningen, Holland 1976.
67. Roggenkamp, R., Sahm, H., Wagner, F.: FEBS Letters 41, 283 (1974).
68. van Dijken, J. P., Veenhuis, M., Vermeulen, C. A., Harder, W.: Arch. Microbiol. 105, 261 (1975).
69. Roggenkamp, R., Sahm, H., Hinkelmann, W., Wagner, F.: Euro. J. Biochem. 59, 231 (1975).
70. Fukui, S., Kawamoto, S., Yasuhara, S., Tanaka, A.: Euro. J. Biochem. 59, 561 (1975).
71. Fujii, T., Tonomura, K.: Agr. Biol. Chem. 36, 2297 (1972).
72. Kato, N., Tani, Y., Ogata, K.: Agr. Biol. Chem. 38, 675 (1975).
73. Sahm, H., Wagner, F.: Arch. Microbiol. 90, 263 (1973).
74. Schütte, H., Floßdorf, J., Sahm, H., Kula, M. R.: Euro. J. Biochem. 62, 151 (1976).
75. Sahm, H.: Arch. Microbiol. 105, 179 (1975).
76. Harder, W., van Dijken, J. P.: In: Microbial Growth on C$_1$-Compounds the society of Fermentation Technology, Japan, 1975, p. 155.
77. Davey, J. F., Whittenbury, R., Wilkinson, J. F.: Arch. Microbiol. 87, 359 (1972).
78. Strom, T., Ferenci, T., Quayle, J. R.: Biochem. J. 144, 465 (1974).
79. Colby, J., Zatman, L. J.: Biochem. J. 148, 513 (1975).
80. Dix, B.: Diplomarbeit, University, Braunschweig 1975.
81. Quayle, J. R.: Adv. Microbial Physiol. 7, 119 (1972).
82. Sahm, H., Wagner, F.: Arch. Microbiol. 97, 163 (1974).
83. Fujii, T., Asada, Y., Tonomura, K.: Agr. Biol. Chem. 38, 1121 (1974).
84. van Dijken, J. P., Harder, W.: Biotechn. Bioeng. 17, 15 (1975).
85. Reuss, M., Sahm, H., Wagner, F.: Chem. Ing. Techn. 46, 669 (1974).
86. Pilat, P., Prokop, A.: Biotechn. Bioeng. 17, 1717 (1975).
87. Reuss, M., Gnieser, J., Reng, H. G., Wagner, F.: Euro. J. Appl. Microbiol. 1, 295 (1975).
88. Payne, W. J.: Ann. Rev. Microbiol. 29, 17 (1970).
89. Abbott, J. B., Laskin, A. I., Meloy, C. J.: J. Appl. Microbiol. 25, 787 (1973).
90. Nemethy, G.: Angewandte Chemie, Internat. Edition 6 (3) 915 (1967).
91. Lücke, J., Oels, U., Schügerl, K.: Chem. Ing. Techn., 48, 573 (1976); 49, 161 (1977).
92. Sonntag, H., Strenge, K.: Koagulation und Stabilität disperser Systeme. Berlin: 1970 VEB Deutscher Verlag der Wissenschaften 1970.
93. Meister, B., Scheele, G. F.: A. I. Ch. E. J. 13, 682 (1967).
94. Rayleigh, Lord: Phil. Mag. (London) 34, 177 (1892).
95. Tyler, E.: Phil. Mag. (London, Edinburg, Dublin) 16, 504 (1933).
96. Hinze, J. O.: A. I. Ch. E. J. 1, 289 (1955).
97. Mateles, R. I.: Biotechn. Bioeng. 13, 581 (1971).
98. Cejka, A.: 3. Symposium Technische Mikrobiologie Berlin 1973, Inst. f. Gärungsgewerbe, S. 281.
99. European Chemical News 15, 3, 30 (1974).
100. MacLaren, D. D.: Chem. Technol. 594 (1975).
101. Schrader, U., Vogelmann, H., Wagner, F.: D. P. application, P 260-4959.8.
102. Koide, K., Hayashi, T., Sumino, K., Iwamoto, S.: Chem. Eng. Sci. 31, 963 (1976).
103. Graham, D. E., Phillips, M. C.: Chapter 16. The conformation of proteins at the air-water interface and their role in stabilizing foams. In: Foams, Int. Symp. Brunel University 1975. Smith, A. L., (Ed.) New York: Acad. Press 1975.
104. Benjamins, J., Feijter, J. A., de Evans, M. T. A., Graham, D. E., Phillips, M. C.: Faraday Discussions of the Chemical Society No. 59, 218 (1975).
105. Jederström, G., Rydhag, L., Friberg, S.: J. Pharmaceutical Sciences 62 (12) 1979 (1973).
106. Saito, H., Friberg, S.: Praman, Suppl. No. 1, 537 (1975).
107. Schügerl, K.: Chem. Ing. Techn. 49, 605 (1977).

Sorption Characteristics for Gas-Liquid Contacting in Mixing Vessels

M. Zlokarnik

Engineering Department of Applied Physics, Bayer AG,
D-5090 Leverkusen, West Germany

Contents

Summary

This paper deals with the determination of absorption rates in mixing vessels for pure water (coalescent conditions) and for aqueous salt solutions (noncoalescent conditions). Two new measuring techniques will be described. The (non-steady-state) Pressure Gauge Method can be used for any pure gas and any liquid. The (steady-state) Hydrazine Method allows measurements in water or in aqueous solutions without changing the physical or chemical properties of the system. The results are evaluated according to the theory of similarity, the dimensionless process numbers being formed from intensively formulated process parameters. Two correlations were thus obtained, one valied for a coalescent and one for a noncoalescent system. The following process characteristics will be introduced: hollow stirrers and injectors in a noncoalescent system; propeller stirrer, hollow stirrer, flat blade turbine, and an injector for a coalescent system. In the case of the flat blade turbine, the parameter liquid height/vessel diameter was varied by the ratio 1:3.

Introduction

It is normal practice to use agitated vessels for the enhancement of mass transfer between a gas and a liquid. This process is generally called gas-liquid contacting and is widely applied in the field of chemical reaction engineering (e.g., for chlorinations, hydrogenations, oxidations, etc.) and in biochemical engineering (incl. biologic waste-water treatment).

This important unit operation has been the subject of many publications which have dealt with the determination of either the interfacial area a, or the liquid-phase mass transfer coefficient k_L, or the absorption rate coefficient $k_L a$. In addition to the oldest works [1, 2], reference should be made to the fundamental papers of Vermeulen et al. [3], Calderbank [4], and Westerterp et al. [5]. In these papers a or k_L was determined by introducing a gas into water (pure liquid) and using physical measuring methods (i.e., gas hold-up [2] or photoelectric measurements of the gas dispersion [3, 4]). On the other hand, the absorption rate coefficient $k_L a$ was always determined by introducing a gas into an aqueous solution and using chemical measuring methods (i.e., oxidation of a sodium sulphite solution, catalyzed by Cu^{2+} or Co^{2+} ions) [1, 5].

Calderbank [4] and Yoshida et al. [6] have already pointed out in their work the significant difference between gas dispersion in a pure liquid and that in a solution. In the first case, coalescent conditions exist in which the fine bubbles produced by the dispersion device coalesce into larger bubbles as soon as they leave the region of high shear stress. In the second case, noncoalescent conditions exist in which, according to the type and concentration of additives (salts or miscible liquids), the coalescence of the primarily produced gas bubbles is more or less hindered, thus greatly increasing the interfacial area. Yoshida [6] explained this by assuming an electrostatic potential of the resultant ions at the liquid side of the interface.

It is known from the work of Langmuir and others that the salts which give negative potentials on the interface (i.e., inorganic salts of strong electrolytes) often increase the surface tension of water [7]. In this case, it follows from the Gibbs adsorption equation, that the net concentration of salt in the interface is less than in the bulk of the liquid. Nevertheless, the Gibbs equation does not satisfactory describe coalescence phenomena [8].

According to [7], it is the specific differences between the anions and cations and their respective interactions with the interface which lead to the formation of the electrical double layers around the gas bubbles. This results in a negative charge on the outside of each gas bubble and causes them to repel each other. This opinion, that the hindering of coalescence is due to the water structure as influenced by the ions and by the interface [7], was convincingly confirmed by Lessard and Zieminski [9]. They found that those salts which most lower the entropy of solution or the self-diffusion of water are required in the lowest concentration to decrease coalescence.

Robinson and Wilke [10] were the first to study systematically the "salt effects" in gas-liquid contacting in mixing vessels. They correlated $k_L a$ values with ionic strength and with the product $(P/V)^n \cdot v_s^m$ (P/V—mixing power per unit volume, v_s—superficial gas velocity). They found for water ($\Gamma = 0$ g ions/l): $n = 0.4$ and $m = 0.35$ and for solutions with $\Gamma \geqslant 0.4$ g ions/l: $n = 0.9$ and $m = 0.39$. In the transition range within these limits a steady trend in the powers was observed.

In a system with coalescent conditions, it is practically impossible to alter the value of k_L by increasing mixing intensity or gas throughput. On the other hand, it was found [11] that under noncoalescent conditions (aqueous electrolyte solutions) the interfacial area per unit volume a can be increased by increasing P/V, whereas k_L and the bubble diameter decrease simultaneously.

This paper deals with experimental work carried out in 1973/74. The original objective was to study the influence of salt addition on the coalescence of gas bubbles in aqueous solutions. For this purpose a new measuring technique was applied which will be described below under the name of the Pressure Gauge Method. It has the advantage over all previous methods that it is system-independent, i.e., it can be applied to any liquid and any pure gas.

However, to obtain reliable rules for scaling-up mixing conditions for required absorption rates in gas-liquid contacting, measurements have to be carried out in vessels of different sizes, possibly including bench and industrial scale. Such a procedure would involve no difficulties when the oxidation of aqueous sodium sulphite solution could be applied as the measuring technique, maintaining steady-state conditions during the absorption process. The data obtained with this technique is only applicable to non-coalescent systems because of the high concentration of sodium sulphite solution that is necessary. To obtain information for coalescent systems, previous measurements had been carried out under non-steady-state conditons (absorption with subsequent desorption of a gas in a pure liquid). To make these tests in larger vessels easier, the Hydrazine Method was developed which permits measurements under steady-state conditons in water with < 1 g salt/l, i.e., in a coalescent system.

Although some of the authors mentioned have evaluated their data in a dimensionless form, a consequent and comprehensive dimensional analysis of the considered process has not as yet been carried out.[1] In this paper this will be done, introducing both process parameters—mixing power and gas throughput—as intensively formulated variables. The resulting complete set of dimensionless numbers and the comprehensive experimental results will enable us to establish two process characteristics for gas-liquid contacting in mixing vessels, one for coalescent, and one for noncoalescent systems. These characteristics have the advantage of being valid for different gas-dispersion devices, as will be proved with different types of stirrers and injectors.

Methods of Measuring Mass Transfer in Gas-Liquid Contacting

Before the two new methods are described, a short review of the techniques usually used will be given.

In general, steady- and non-steady-state techniques can be employed with regard to whether or not the concentration c of the gas being dissolved in the liquid remains constant.

Steady-state methods have the advantage of greater accuracy and give results which are easier to evaluate. For the maintenance of steady-state conditions ($c = $ const) a chemical reaction is commonly used to remove the absorbed gas. This reaction must be so fast that the mass transfer is the rate-controlling step. This is the case with the oxidation of aqueous sodium sulphite solution with air or oxygen and the reaction of sodium hydro-

---

[1] Yagi and Yoshida [12] have recently correlated the results of their measurements with a set of dimensionless numbers. The measurements were carried out on aqueous solutions of glycerol, CMC, and PAN, allowing higher liquid viscosities to be investigated. In the range of higher viscosities the coalescent system seems to prevail.

xide with CO_2 [13]. The sulphite oxidation has to be accelerated catalytically. When Cu^{2+} ions are used, the oxidation is so fast that the absorption takes place at $c = 0 =$ const. When Co^{2+} ions are used the reaction is even faster, so that it does not take place in the bulk of the liquid but at the interface [14]. Absorption accelerated in this way is called chemisorption. The disadvantages of chemical methods include a high heat of reaction and a relatively high salt concentration which always results in a noncoalescent system.

The Hydrazine Method (which will be fully described below) is a stationary method without either of these disadvantages. The process takes place at $c = $ const $\neq 0$ and a salt concentration of < 1 g/l, i.e., in a coalescent system. Nevertheless, due to $c \neq 0$ the O_2 concentration in the liquid has to be measured with an oxygen membrane electrode. In *non-steady-state methods* the absorption rate is calculated from the change in the concentration of the gas dissolved in the liquid with time. Previously, for this purpose, the gas concentration had to be continuously measured in the liquid or in the gas before and after it had passed through the liquid. The non-steady-state Pressure Gauge Method, presented in this paper, seems to be far easier and more accurate than the above-mentioned, especially for very fast absorption processes.

The *Pressure Gauge Method* will be described in connection with the apparatus (Fig. 1) used in this work. It essentially consists of a stirring vessel (1) and two cylinders for the gas: the gas supply cylinder (2) and the measuring cylinder (3). The stirring vessel has a

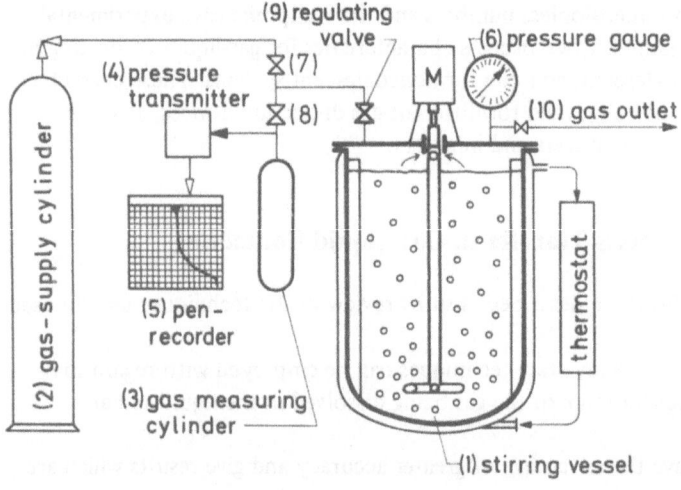

Fig. 1. Apparatus for the measurement of absorption rates by the Pressure Gauge Method

water jacket which is connected to a thermostat. The dimensions of the stirring vessel are: diameter 0.40 m, height 0.60 m. The vessel is fully baffled with four baffles (width 0.1 diameter of the vessel). A self-aspirating hollow stirrer [15a], attached to a hollow shaft, is used. This stirrer sucks the gas from the gas space above the liquid and disperses it into the liquid volume, thus an additional gas circulating pump is not required. The hollow stirrer used here is a very efficient tube stirrer with a diameter of 0.06 m and a bottom clearance of 0.10 m (see [16] for details). The stirrer shaft is sealed by a rubber lip seal.

The procedure is as follows: the stirring vessel is filled with exactly 60 l of the liquid. The liquid is intensively stirred until it is at $20 \pm 0.5\,°C$. During this time a small gas throughput is maintained through the gas space of the stirring vessel [valves (7), (9), and (10) open]. In all experiments using this method, oxygen of 99.9% purity was used. After approx. 20 min of stirring, the liquid is saturated with O_2 at 20 °C and 1 bar of oxygen. Now the stirrer is stopped and valve (10) closed. Valve (7) is then fully opened to allow a rapid rise of O_2 pressure up to 2.0 bar in the gas space. Then valves (7) and (9) are closed and valve (8) opened. Now the actual measurement can begin. When the stirrer is started, the gas is absorbed by the liquid according to the higher system pressure which results in a pressure drop in the gas space. To prevent this, gas losses have to be compensated from the measuring cylinder (3) using the regulating valve (9), thus keeping the absorption at 2 bar as indicated by pressure gauge (6). The resulting pressure drop in the measuring cylinder (3) is transmitted (4) and registered by a pen-recorder (5). The measurement is finished when saturation at 2 bar is achieved. Then the pressure in the gas space is slowly released to 1 bar while stirring and then the measurement can begin again under different stirring conditions. The rotational speed of the stirrer was varied between 1000 and $2400\ \text{min}^{-1}$, thus simultaneously varying the power input and the gas throughput.

In order to evaluate these measurements one has to consider that the pressure p in the measuring bomb at the beginning of the experiment ($t = 0$) corresponds to the saturation concentration c_s at 1 bar, the pressure during the measurement p (t) corresponds to c (t) and the pressure at the end of the measurement p $(t = \infty)$ corresponds to the saturation concentration c_s at 2 bar. When the expression $[p(t = \infty) - p(t)]/p(t = \infty)$ is plotted against time t on log-linear paper, a straight line is produced, the slope of which is $k_L a$.

Knowing the exact volume of the measuring cylinder (3) including the pipe volume until valve (9) and the density of oxygen under measuring conditions, the amount of dissolved oxygen in 60 l of liquid can be calculated and from this the solubility of the gas in the liquid involved can be evaluated.

The *Hydrazine Method* takes advantage of the fact that the reaction between hydrazine and oxygen according to the relationship

$$N_2H_4 + O_2 \rightarrow N_2 + H_2O$$

gives no reaction products which would change the physical or chemical properties of the system. The most favorable conditions for this reaction are at pH = 11–12 and a $CuSO_4$ concentration of 0.01 M. The procedure is as follows: the vessel is filled with water and the required amount of $CuSO_4$ is dissolved by stirring. Sodium hydroxide is added until the indicated pH is reached, also producing $Cu(OH)_2$. To prevent a reduction of $Cu(OH)_2$ to Cu_2O during the following hydrazine addition, a certain O_2 level (3–4 ppm) must be achieved by introducing air or O_2 before the test begins. After the absorption conditions have been set up, an appropriate hydrazine flow into the system is maintained. This maintains the O_2 concentration in the liquid at a constant value between

1 ppm and c_s. The concentration of the aqueous hydrazine solution supplied to the system depends on the intensity of absorption and on the accuracy of the pump used. The O_2 concentration in the system is measured by an oxygen membrane electrode. Under steady-state conditions the absorption rate equals the reaction rate. The amount of hydrazine introduced per unit time into the system equals the amount of oxygen absorbed per unit time due to the equal molecular masses of N_2H_4 and O_2. Mass balances proved that salt concentration in the liquid never rose above 1 g/l. Therefore the system was always in a coalescent condition.

After the measurements have been completed, the liquid is aerated to a higher oxygen level to prevent reduction of $Cu(OH)_2$ by traces of hydrazine. The liquid can be used again.

Consideration of the Absorption Process from the Point of View of the Theory of Similarity

A consideration of the physical aspects of the absorption process usually begins with the relationship

$$G/V = k_L a\, \Delta c \tag{1}$$

which is based on the two-film theory, where G/V is the mass throughput per unit volume of liquid, k_L is the liquid-phase mass transfer coefficient, a is the gas-liquid interfacial area per unit volume of liquid and Δc is a characteristic concentration difference. This relationship (1) is based on the following assumptions:

a) The intensity of the gas-liquid contacting is so high that a quasi-uniform system is produced. This means that in every element of liquid the bubble density and, therefore, the interfacial area is the same.

b) The gas-phase mass transfer coefficient k_G is so high in comparison to k_L that it can be neglected (according to [4] $k_G/k_L > 50$).

c) The absorption rate at the interface is extremely fast, resulting in an equilibrium concentration of the dissolved gas at the interface $c^* = c_S\,(\vartheta, x, p)$ and therefore $\Delta c = c_s - c$. In accordance with (a), Δc must be formulated to take into account the quasi-uniformity of the system. Because of efficient back-mixing, the concentration of the gas dissolved in the liquid is constant (c is position-independent), whereas the c_s at the interface of the bubbles is position-dependent due to the changing hydrostatic pressure in the vessel. It was found that the absorption rates of numerous measurements in mixing vessels and in bubble columns can be correlated best with logarithmic mean concentration difference Δc_m, based on the segregation of the gas-phase, even when the conditions were actually coalescent:

$$\Delta c_m \equiv (c' - c'')/\ln\left[(c' - c)/(c'' - c)\right], \tag{2}$$

$$c' = c_s\, x' p', \tag{3}$$

$$c'' = c_s\, x'' p'', \tag{4}$$

$$x'' = (qx' - G/\rho_G)/(q - G/\rho_G). \tag{5}$$

In these equations c', c''; x', x'', and p', p'' are the saturation concentrations, mol fractions and system pressures of the gas respectively at the inlet (') and the outlet (") of the system. q is the gas throughput and ρ_G the gas density under standard conditions. In the general consideration of the mass transfer, k_L and a are combined to give the absorption rate coefficient $k_L a$ which is defined by (1) as

$$k_L a = G/(V \Delta c_m). \tag{6}$$

Because of the basic equality of absorption and desorption processes, the term *sorption rate coefficient* will be used for $k_L a$ in the following text and the corresponding process characteristics will be named *sorption characteristics*.

According to the assumptions (a)–(c) on which relationship (1) is based, the following have to be taken into account when a list of parameters influencing the main parameter $k_L a$ is drawn up:

(d) $k_L a$ must be independent of all geometrical parameters (i.e., diameters of the stirrer and the tank, etc.) because of the assumption of a quasi-uniform system.

(e) $k_L a$ must be independent of the material parameters of the gas-phase (i.e., gas density, gas viscosity) because of (b).

(f) $k_L a$ is an intensive quantity because of its volume-related formulation in (1). According to this, all process parameters have also to be formulated as intensive quantities.

The relevant material parameters of the liquid-phase to be considered are:

ρ — liquid density

ν — liquid kinematic viscosity

σ — liquid surface tension

D — diffusivity of the gas in the liquid

S_i — material parameters which describe coalescence behavior of solutions (i.e., ionic strength Γ, electrical charge of the ions, etc.). At the moment, it is not definitely known which and how many quantities describe this phenomenon quantitatively (e.g., the behavior of aqueous alcohol solutions cannot be described by parameters mentioned in [8–10]).

The process parameters are:

P/q power for the dispersion per unit gas throughput

q/V gas throughput per unit volume of liquid

g gravitational constant (because of the large difference between gas and liquid densities).

Thus for $k_L a$ the following functional dependence results:

$$k_L a = f(\rho, \nu, \sigma, D, S_i, P/q, q/V, g). \tag{7}$$

By means of dimensional analysis, (7) can be reduced to a dependence between dimensionless numbers [18]:

$$(k_L a)^* = f_1\{(P/q)^*, (q/V)^*, \sigma^*, Sc, S_i^*\}. \tag{8}$$

The symbols in brackets with asterisks represent the following dimensionless groups:

$(k_La)^* \equiv k_La\,(\nu/g^2)^{1/3}$

$(P/q)^* \equiv (P/q)\,[\rho(\nu g)^{2/3}]^{-1}$

$(q/V)^* \equiv (q/V)\,(\nu/g^2)^{1/3}$

$\sigma^* \equiv \sigma[\rho(\nu^4 g)^{1/3}]^{-1}$

$Sc \quad\ \equiv \nu/\mathbb{D}$ (Schmidt number)

S_i^*.

It follows from (8) that the main number $(k_La)^*$ depends on two process numbers $(P/q)^*$ and $(q/V)^*$ and on at least three material numbers σ^*, Sc, and S_i^* (when $i = 1$). (Although the dimensions of S_i are not known, it will always be possible to transform them with the aid of ρ, ν, and g into pure material numbers.)

However, the dependence (8) is all that can be contributed by the theory of similarity. The function f_1 has to be determined experimentally. Since the experimental data were obtained from measurements on either water or 1N sodium sulphite aqueous solutions, the numbers σ^* and the Schmidt number Sc remained practically constant, but S_i^* changed from values for coalescent conditions to values for highly noncoalescent conditions. These experimental data will, therefore, allow the evaluation of the process characteristics for these two systems.

Evaluation and Discussion of Experimental Results

Influence of Salt Concentration on k_La

Figure 2 essentially shows the experimental results obtained with the Pressure Gauge Method on water and aqueous solutions of Na_2SO_4 at different concentrations. Seven measurements were carried out on each system, increasing the rotational velocity from 1200 to 2400 min^{-1} in increments of 200 min^{-1}. The results are given in the form of $k_La(P/V)$, although it should be remembered that increasing the rotational velocity of hollow stirrers enhances the gas throughput as well as the power. Following the trend of each set of measurements, the proportionality $k_La \propto P/V$ can easily be seen.

The conspicuous results of these measurements is the surprisingly high dependence of k_La on the salt concentration. For a 1.0N solution of Na_2SO_4, containing 71 g salt/l, the k_La values are six times higher than for pure water. This is due not only to the previously mentioned "salt effect", but also to the fact that a hollow stirrer which produces very fine gas bubbles was used.

Using the Pressure Gauge Method, in addition to the absorption rates, the gas solubilities were measured. In Table 1 these solubility values are listed for the systems O_2/water and O_2/aqueous Na_2SO_4 solutions as well as for the systems N_2/1.0N Na_2SO_4 and N_2/1.0N Na_2SO_3. Values for the latter two systems were determined to obtain information about the solubility of O_2 in aqueous Na_2SO_3 solutions. Since no difference was found between the solubility of N_2 in 1.0N Na_2SO_4 and in 1.0N Na_2SO_3, the same will be postulated for the solubility of O_2 in these two solutions. This information was necessary to evaluate the measurements of the sodium sulphite oxidations in the same way as the others.

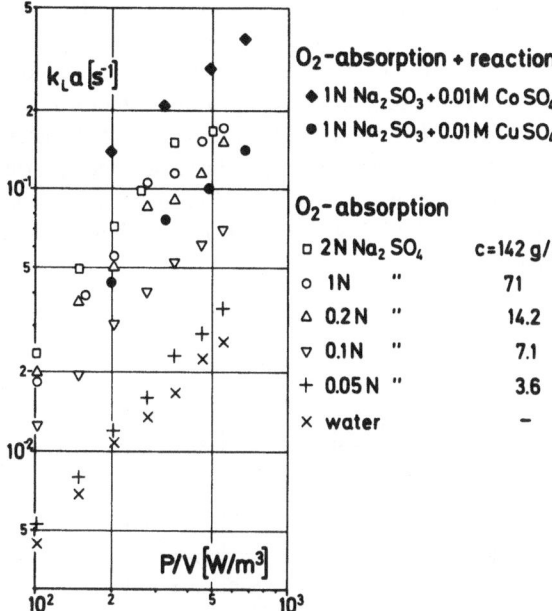

$k_L a$ [s⁻¹]

O₂-absorption + reaction

♦ 1N Na₂SO₃+0.01M Co SO₄

● 1N Na₂SO₃+0.01M Cu SO₄

O₂-absorption

□ 2N Na₂ SO₄	c=142 g/l
○ 1N "	71
△ 0.2N "	14.2
▽ 0.1N "	7.1
+ 0.05N "	3.6
× water	–

P/V [W/m³]

Fig. 2. $k_L a$ (P/V) dependences for different aqueous solutions of salts. Physical absorption was measured here by the Pressure Gauge Method

Table 1. Solubility of O_2 and N_2 in different liquids

Gas	Liquid	$10^2 c_s$ [kg (gas)/m³_L bar]
O_2	Water	4.1
	0.1N Na₂SO₄	4.0
	1.0N Na₂SO₄	2.8
	2.0N Na₂SO₄	1.8
	0.1N NaCl	4.1
	1.0N NaCl	3.0
N_2	Water	2.0
	1.0N Na₂SO₃	1.2
	1.0N Na₂SO₄	1.2

In Fig. 2, using the assumed solubility of O_2 in 1.0N Na₂SO₃ solution, the results of the measurements from the sulphite method are also plotted. The expected agreement between the physical absorption and Cu^{2+} catalyzed sulphite oxidation is not very good. The main reason is the different evaluation of a steady-state and a non-steady-state process with regard to the characteristic concentration difference: $\Delta c_m \neq \Delta c$.

A comparison between Cu^2- and Co^{2+}-catalyzed oxidations shows that the latter is three times faster: this proves that chemisorption takes place where the reaction front advances from the bulk of the liquid to the interface.

When the dependences $k_L a(P/V)$ for each solution are related to that of water, an enhancement factor m is found which is a function of S_i parameters of the system. In Fig. 3 m is plotted against salt concentration (and not for instance against ionic strength) for the sole reason that m for aqueous methanol solutions could then also be plotted. It was found that methanol (and also ethanol and acetone) produces a far greater hindrance to coalescence than does the same concentration of a strong electrolyte. Furthermore, it can be seen that a marked influence of salt concentration on the system behavior begins at 5 g/l, it greatly increases until 20 g/l, and still has not reached its maximum at 100 g/l.

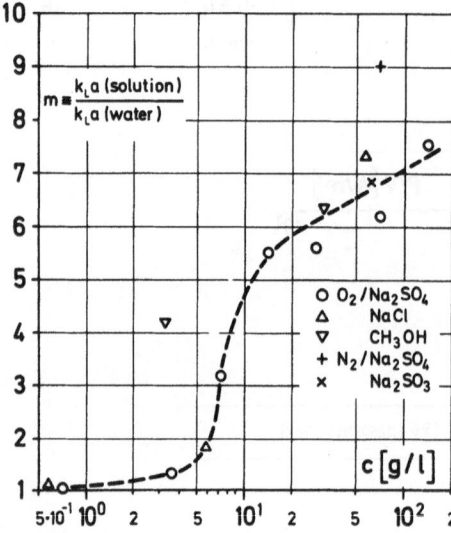

Fig. 3. Enhancement factor m as a function of the salt concentration c

In Fig. 2 two sets of $k_L a$ values were shown which were obtained using the sulphite method. It is worth mentioning that another interesting difference between Co^{2+}- and Cu^{2+}-catalyzed sulphite oxidation exists, which seems to have been previously overlooked. Figure 4 shows the $k_L a \, (P/V)$ dependences obtained by using the same apparatus. For the Cu^{2+}-catalyzed reaction the same results were obtained when either air or oxygen was used. On the other hand, a relatively large difference was found for the Co^{2+}-catalyzed reaction. The reason for this behavior can easily be proved by calculation and is due to the following: in the Co^{2+}-catalyzed chemisorption in the case of O_2, the absorption is so fast that the gas bubbles shrink and so the interfacial area of the individual bubble decreases during the residence time. This does not occur with air because the great ballast of nitrogen reduces Δc_m (and therefore G) by a factor of 5. Thus the Co^{2+}-catalyzed sulphite reaction for the case of O_2 does not give the relationships $k_L a(P/V)$ and therefore $a(P/V)$ correctly, because in this case a is not only dependent on P/V but also on the absorption rate G.

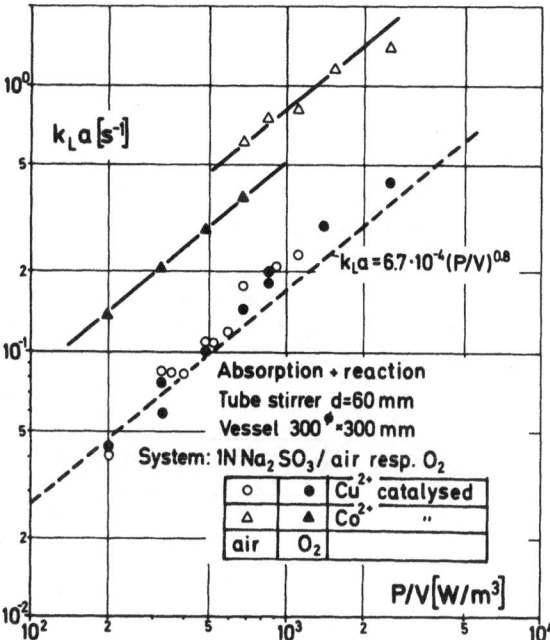

Fig. 4. $k_L a$ (P/V) dependences as measured by the Cu^{2+} and Co^{2+} catalyzed sulphite oxidations with air or oxygen

Noncoalescent System

With the assumed c_s value for O_2 in 1.0N sodium sulphite solution, it is possible to evaluate our own numerous experimental data [14 b, 16], based on the Cu^{2+}-catalyzed sulphite method, according to the dependence (8). This data was obtained with tanks of different shapes and sizes using two different types of hollow stirrer. The so-called three-edge stirrer [14 a] was installed in a baffled vessel of $H/D = 1$. The size of this apparatus was scaled up by factors of 1:2:3:4:5 which were all geometrically similar. The volumes of these devices ranged between 6.3 and 780 l. In the case of the "tube stirrers", a shallow rectangular vessel (a pool) of cross-section 0.8 × 1.0 m and a liquid height of 0.11 m was used. As already mentioned, a change in the rotational velocity of hollow stirrers alters P and q simultaneously. Consequently, for different diameters of these stirrers, different q/V values will result for P/V = const. In our case for a scale-up of 1:5 at a given P/V, the q/V values decreased by up to factor of 6. Nevertheless, no influence of q/V on $k_L a$ could be found. (Obviously the minimum q/V value was sufficient to provide the system with oxygen.)

Because of this fact, the dimensionless numbers $(P/q)^*$ and $(q/V)^*$ have to be combined to a new dimensionless number which does not include q:

$$(P/q)^* \cdot (q/V)^* = (P/V) \ [\rho \ (\nu g^4)^{1/3}]^{-1} \equiv (P/V)^*.$$

The experimental data can now be correlated in the form

$$(k_L a)^* = f_2 \{(P/V)^*\}; \ \sigma^* = const, S_i^* \stackrel{\wedge}{=} noncoalescence,$$

as shown in Fig. 5.

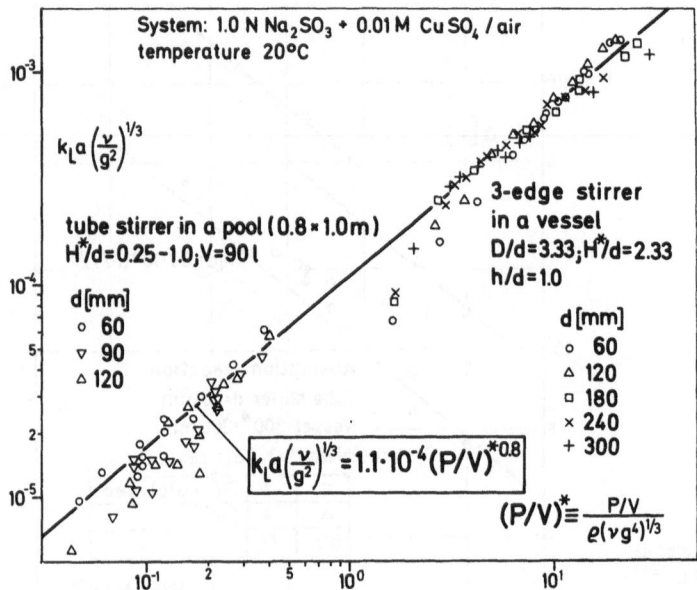

Fig. 5. Sorption characteristic for hollow stirrers in a noncoalescent system. $\sigma^* = 3300$; $Sc = 435$; $S_i^* \hat{=}$ noncoalescent system

From this, it can be seen that there are some slight differences in the function f_2 for vessels of different sizes.

This can be clearly seen from the data for the three-edge stirrer by plotting straight lines through the points for respective diameters of the stirrer. The results obtained with the diameter $d = 60$ mm ($D = 200$ mm, $V = 6.3$ l) give a higher dependence $(k_L a)^* = f\{(P/V)^*\}$ than the results obtained with $d = 120$ mm ($D = 400$ mm, $V = 50$ l). Only when the diameter reaches $d = 180$ mm ($D = 600$ mm, $V = 170$ l) does this function become scale-independent. The reason for this is the presence of the liquid surface where mass transfer also occurs. This becomes insignificant in larger vessels. This behavior explains the discrepancy in results obtained by different researchers. To obtain reliable data for scale-up in gas-liquid contacting, the smallest vessel diameter should be 500 mm.

The straight line in Fig. 5 describes the whole experimental data with sufficient accuracy over the whole range of $(P/V)^*$ values. Thus, the *sorption characteristic* of a mixing vessel with a hollow stirrer for a noncoalescent system is given by the relationship

$$(k_L a)^* = 1.1 \cdot 10^{-4}\{(P/V)^*\}^{0.8}; 5 \cdot 10^{-2} < (P/V)^* < 2 \cdot 10^1. \tag{9}$$

According to Eq. (1) this sorption characteristic can also be expressed by

$$G/\Delta c_m \propto P^{0.8} \cdot V^{0.2}. \tag{9a}$$

This sorption characteristic will also be valid for a flat blade turbine with the gas supply from below, because this type of stirrer also produces a region of high shear stress which is necessary for the generation of fine bubbles. Contrary to this, the propeller stirrer is an axial pump which causes little shear stress and so fine bubbles are not produced.
In Fig. 6 the sorption characteristic for an injector is compared with that for stirrers. The injector used was similar to the Penberthy XL-96 (for details see [21]) with a water nozzle diameter of 8 mm and a diffusor throat diameter of 14 mm. The injector was positioned half-way up the wall and was inclined downwards at $45°$. The vessel used was $1.0 \phi \times 1.3$ m. It can be seen from Fig. 6 that the relationship $(k_L a)^* \propto \{(P/V)^*\}^{0.8}$ still holds, but the injector is 2.5 times more efficient than the stirrer. The reason for this is that the injector uses a greater proportion of the power[2] for the dispersion of the gas into the liquid than a stirrer which chiefly mixes the liquid.

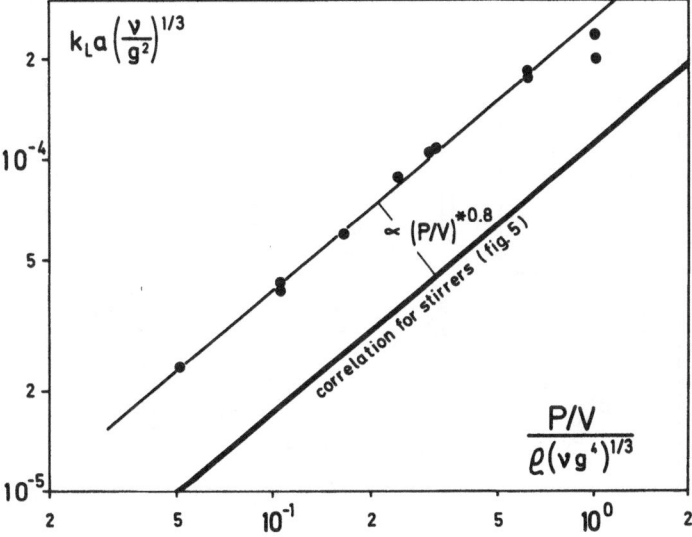

Fig. 6. Comparison between the sorption characteristics for an injector and for hollow stirrers. In the case of the injector, P_L is calculated from the power of the liquid jet[2].

This finding, that the number $(q/V)^*$ has practically no influence on the sorption characteristic for noncoalescent systems, is due to the fact that fine bubbles are conserved in this system; they have only a weak tendency to escape from the liquid and willingly follow the liquid patterns. Thus their hold-up is only dependent on the degree of turbulence and therefore only on P/V.

---

[2] The power P_L of the liquid jet in an injector is calculated from the pressure drop Δp_L of the liquid in the nozzle and the liquid throughput q_L:

$$P_L = \Delta p_L \cdot q_L.$$

Coalescent System

The experimental results presented here have generally been obtained by the Hydrazine Method. In order to obtain independent variation of power and gas throughput, flat blade turbines with the gas supply from below were used.

Figure 7 shows the results obtained in a vessel (400 $\phi \times$ 400 mm) with a flat blade turbine of 90 mm diameter. It can be seen that the number $(q/V)^*$ has a great influence on absorption. The correlation of this data can be achieved by combining the numbers $(k_L a)^*$ and $(P/V)^*$ with $(q/V)^*$ so that the parameter V in both numbers is cancelled:

$$(k_L a)^* \cdot (q/V)^{*-1} = G/(q \Delta c_m)$$

$$(P/V)^* \cdot (q/V)^{*-1} = (P/q) \left[\rho (vg)^{2/3} \right]^{-1} \equiv (P/q)^*.$$

In this way the second number takes the form suggested in Eq. (8).

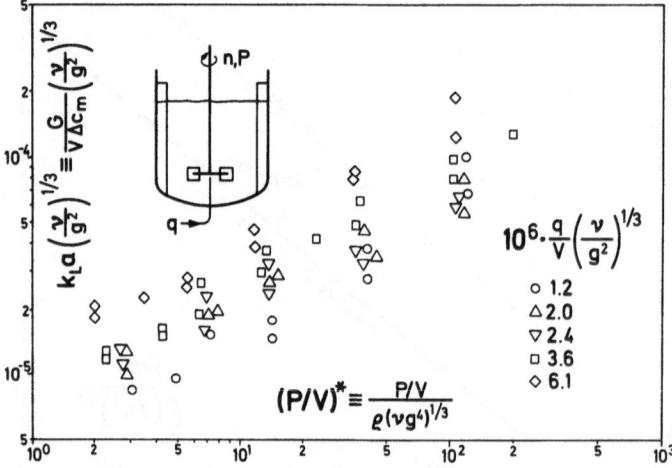

Fig. 7. Proof of the influence of $(q/V)^*$ on $(k_L a)^*$ in a coalescent system. Vessel 400 $\phi \times$ 400 mm, flat blade turbine d = 90 mm, Hydrazine Method

Figure 8 shows the experimental data from Fig. 7 using the above numbers for the correlation. Measurements of non-steady-state absorption with a tube stirrer of 90 mm diameter and measurements of a steady-state absorption with a larger flat blade turbine (d = 150 mm) are included in this figure. All the data is described by the process equation

$$G/(q \Delta c_m) = 0.015 (P/q)^{*0.5}; 3 \cdot 10^4 < (P/q)^* < 1 \cdot 10^7 \tag{10}$$

which can be expressed by

$$G/\Delta c_m \propto (P \cdot q)^{0.5}. \tag{10a}$$

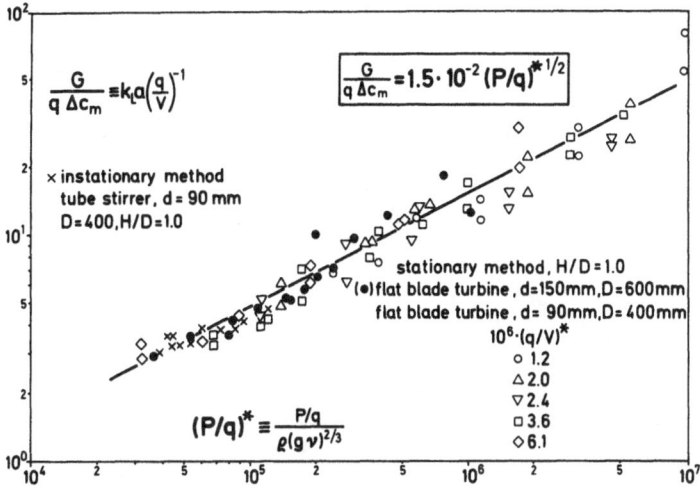

Fig. 8. Sorption characteristic for hollow stirrers and flat blade turbines in a coalescent system. $\sigma^* = 3\,300$; $Sc = 435$; $S_i^* \hat{=}$ coalescence system

The finding that the gas throughput has a distinct influence on the absorption rate in a coalescent system can be explained in this case by the presence of relatively large bubbles, which are formed by coalescence from the fine primary bubbles generated by the stirrer. The large bubbles have a greater buoyancy and escape quickly from the liquid. The gas hold-up has to be maintained by the gas throughput.

According to the fact that in a coalescent system fine primary bubbles coalesce very quickly to large bubbles, it can be assumed here that the dispersion device has little influence on the sorption characteristic. This assumption has been verified, as shown in Fig. 9. The circles represent the measurements with a propeller stirrer of 290 mm diameter in a vessel of $1.0 \phi \times 1.0$ m, the air throughput being supplied through a perforated pipe ring (diameter 218 mm) from below. The filled circles give the results of an injector and vessel as described in Fig. 6. The straight line is the same as in Fig. 8, which represents the results obtained with a flat blade turbine. It can be seen from Fig. 9 that at low values of $(P/q)^*$ the data for all three different dispersion devices are practically the same, whereas at high values of $(P/q)^*$ they differ considerably. This is due to different liquid patterns in the vessel; the propeller stirrer as an axial pump produces a circulation which holds the gas bubbles in the liquid better than the circulation produced by a flat blade turbine. The injector produces feeble circulation, so that gas bubbles escape easily.

We conclude from Fig. 9 that the liquid pattern has a large influence on the gas hold-up and therefore on the interfacial area in coalescent systems. Now the question has to be posed, how the height/diameter ratio H/D influences the liquid pattern and consequently the bubble hold-up and the absorption rate. This was investigated in a vessel of 600 mm diameter with a flat blade turbine of 150 mm diameter. The H/D ratio was

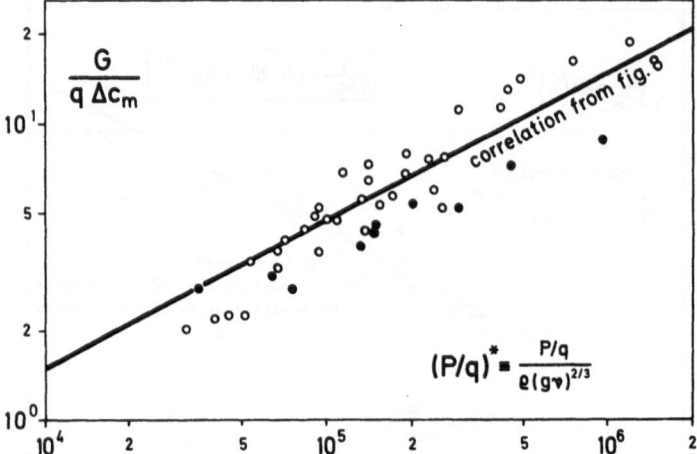

Fig. 9. Comparison of different types of dispersion devices for gas-liquid contacting in a coalescent system. *Straight line:* flat blade turbine (Fig. 8). *Circles:* propeller stirrer (vessel 1.0 ϕ × 1.0 m; d = 270 mm). *Full circles:* injector (vessel 1.0 ϕ × 1.3 m; see also Fig. 6)

varied between 1.0 and 3.0. The results of these measurements are shown in Fig. 10. It can be seen that for $H/D < 2.0$, it is the stirrer which governs the liquid behavior in the vessel. Above this value the influence of the stirrer diminishes and in the range of small $(P/q)^*$ values the flow pattern is produced from the bubble swarm (airlift): the mixing vessel becomes a bubble column with a stirrer as a gas distributor. In the range of $(P/q)^* > 5 \cdot 10^5$, it is again the stirrer which dominates the liquid pattern.

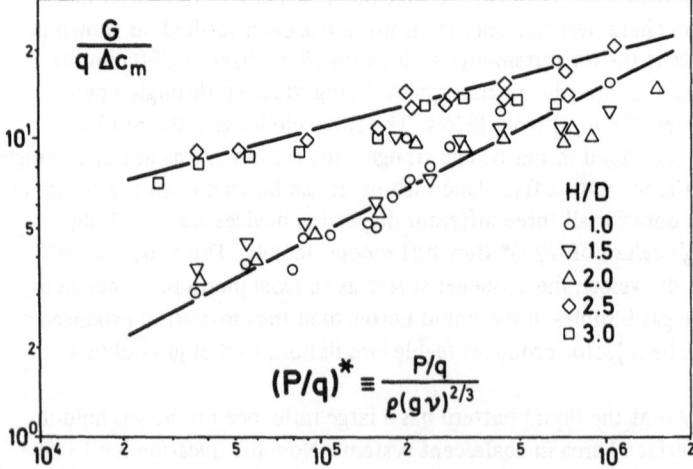

Fig. 10. Sorption characteristic of a flat blade turbine in a coalescent system. Vessel D = 600 mm, stirrer diameter d = 150 mm. H/D = 1–2: $G/(q\Delta c_m)$ = 0.015 $(P/q)^{*0.5}$ (see Fig. 8). H/D = 2.5–3: $G/(q\Delta c_m)$ = 0.60 $(P/q)^{*0.25}$

The sorption characteristics for noncoalescent systems—Eq. (9) and Figs. 5 and 6—and for coalescent systems—Eq. (10) and Figs. 8–10—allow the calculation of the absorption rates for given conditions when water or aqueous solutions have to be contacted with gas in a mixing vessel. For these calculations, the power of the stirrer in the gas/liquid dispersion has to be known in advance. It can be calculated for different types of hollow stirrers from [15a, 16] and for flat blade turbines from [19]. In the case of propeller stirrers, it was surprisingly found that the gas throughput does not influence the mixing power. In this case, the Newton or power number $Ne \equiv P/(n^3 d^5 \rho) = 0.35 =$ const, can also be used for gas-liquid contacting.

The q values in both numbers have to represent the throughputs under the respective conditions: in $(P/q)^*$ it is the effective gas throughput q_1 under the hydrostatic pressure at the level of the stirrer:

$$q_1 = q_N/[1 + 0.1 (H - h)].$$

In $G/(q \cdot \Delta c_m)$ it is the average effective gas throughput q_2 in the vessel:

$$q_2 = q_N/(1 + 0.05 H).$$

Conclusions

The results from the literature quoted and investigations reported in this paper allow the determination of the absorption rates in gas-liquid contacting in mixing vessels for either pure water or an aqueous solution. Furthermore, some of these papers [12, 13] permit the calculation of mass transfer from a gas phase into a system of higher viscosity. However, for gas-liquid contacting in the range of liquid viscosities of $\nu = 10^{-4}$–10^{-3} m^2/s the high-speed stirrers such as propeller stirrers or flat blade turbines are not very suitable. For these systems low-speed stirrers such as frame or paddle stirrers should be used. It is a task for further research to determine the sorption characteristics for these types of stirrers in connection with highly viscous, non-Newtonian and elastoviscous liquids.

Acknowledgements

The essential results of this paper were first presented at the internal meeting of the section "Technical reaction engineering" on 28th February 1974 in Königstein/Taunus and also at the annual meeting of the German Process Engineers on 19th September 1974 in Munich. A summary of this was also published in German [20].

Nomenclature

a [m^{-1}]	gas-liquid interfacial area per unit volume of liquid
c [ppm or kg m^{-3}]	concentration of gas dissolved in the liquid
c [g l^{-1}]	salt concentration of a solution
c_s [ppm or kg m^{-3}]	saturation concentration of the gas dissolved in the liquid
Δc [ppm or kg m^{-3}]	concentration difference
Δc_m [ppm or kg m^{-3}]	log mean concentration difference
d [mm or m]	stirrer diameter
D [mm or m]	vessel diameter
g [m s^{-2}]	gravitational constant
G [kg s^{-1}]	gas throughput through the interface
h [mm or m]	bottom clearance of the stirrer
H [mm or m]	liquid height in the vessel
H^* [mm or m] $= (H - h)$	liquid height above the stirrer
k_L [m s^{-1}]	liquid-phase mass transfer coefficient
$k_L a$ [s^{-1}]	(ab)sorption rate coefficient
n [min^{-1} or s^{-1}]	rotational velocity of the stirrer
P [W or kW]	mixing power in the gas-liquid dispersion
p [bar]	system pressure
q [m^3 s^{-1}]	gas throughput
v_s [m s^{-1}]	superficial gas velocity
V [l or m^3]	liquid volume
x [—]	mol fraction of the absorbed gas in the gas mixture
Γ [g ions l^{-1}]	ionic strength of the electrolyte
ϑ [°C]	system temperature
$\mathbb{D}$ [m^2 s^{-1}]	diffusivity of the gas in the liquid
ρ [kg m^{-3}]	liquid density
ν [m^2 s^{-1}]	liquid kinematic viscosity
σ [kg s^{-2}]	liquid surface tension

All dimensionless numbers are formed by parameters of consistent dimensions.

References

1. Cooper, C. M., Fernstrom, G. A., Miller, S. A.: Ind. Eng. Chem. 36, 504 (1944).
2. Foust, H. C., Mack, D. E., Rushton, J. H.: Ind. Eng. Chem. 36, 517 (1944).
3. Vermeulen, T., Williams, G. M., Langlois, G. E.: Chem. Eng. Progress 51, 85-F (1955).
4. Calderbank, P. H.: a) Trans. Instn. Chem. Engrs. 36, 443 (1958); b) Trans. Instn. Chem. Engrs. 37, 173 (1959).
5. Westerterp, K. R., van Dierendonck, L. L., de Kraa, J. A.: Chem. Eng. Sci. 18, 157 (1963).
6. Yoshida, F. et al.: Ind. Eng. Chem. 52, 435 (1960).
7. In: Recent Progress in Surface Science, Vol. 1. D. A. Haydon, p. 111, New York–London: Academic Press 1964.
8. Marucci, G., Nicodemo, L.: Chem. Eng. Sci. 22, 1257 (1967).
9. Lessard, R. R., Zieminski, S. A.: Ind. Eng. Chem. Fundam. 10, 260 (1971).
10. Robinson, C. W., Wilke, C. R.: Biotechn. and Bioengng. 15, 755 (1973).
11. Robinson, C. W., Wilke, C. R.: AIChE Journal 20, 285 (1974).
12. Yagi, H., Yoshida, F.: Ind. Eng. Chem., Process Des. Dev. 14, 488 (1975).
13. Yoshida, F., Miura, Y.: I & EC Process Des. Dev. 2, 263 (1963).
14. Linek, V.: Chem. Eng. Sci. 21, 777 (1966).
15. Zlokarnik, M.: a) Chem. Ing. Techn. 38, 357 (1966); b) Chem. Ing. Techn. 38, 717 (1966).
16. Zlokarnik, M., Judat, H.: Chem. Ing. Techn. 39, 1163 (1967).

17. Zlokarnik, M.: Chem. Ing. Techn. **42**, 1310 (1970).
18. Zlokarnik, M.: Theory of similarity in process engineering (in German), p. 84, Bayer AG, Leverkusen 1974.
19. Zlokarnik, M.: Chem. Ing. Techn. **45**, 689 (1973).
20. Zlokarnik, M.: Chem. Ing. Techn. 47, 281 (1975).
21. Jackson, M. L.: A.E.Ch.E. Journal **10**, 836 (1964).

Advances in Biochemical Engineering

Editors: T.K. Ghose, A. Fiechter,
N. Blakebrough
Managing Editor: A. Fiechter

Volume 6

New Substrates

28 figures, 26 tables. V, 127 pages. 1977
ISBN 3-540-08363-4

Contents
A. E. Torma: *The Role of Thiobacillus ferro-oxidans in Hydrometallurgical Processes*
The microbiological leaching of metal sulfides
is now a common practice throughout the world
for recovery of copper and uranium from
low-grade ore bearing materials. The present
article critically reviews and regroups the
available data. It points our further research
areas and future possibilities for biogenic action
in hydrometallurgical operations.
(363 references)

T.K. Ghose: *Cellulase Biosynthesis and Hydro-lysis of Cellulosic Substances.*
Cellulosics may become a source energy and
food products. Economic and chemical aspects
are discussed. (100 references)

H. Sahm: *Metabolism of Methanol by Yeasts.*
As a consequence of the industrial interest in
methanol utilizing microorganlisms many
studies on the isolation identification and
physiology of those organisms were made.
(124 references)

J.F. Martin: *Control of Antibiotic Synthesis by Phosphate*
Present evidence suggests that ATP may be the
effector controlling expression of antibiotic
synthesis, although highly pholated
adenine of guanine nucleotides or even cyclic
nucleotides can not be excluded. The final
answer to these questions requires extensive
research of the antibiotic synthesizing enzymes
and of the regulatory mechanisms controlling
the formation and the activity of these enzymes.
(155 references)

Volume 7

Biotechnology

112 figures. VII, 150 pages. 1977
ISBN 3-540-08397-9

Contents
K. Schügerl, J. Lücke, U. Oels: *Bubble Column Bioreactors – Tower Bioreactors Without Mechanical Agitation*
The application of bubble column fermentors in
biotechnology is hindered by lack of the
necessary experimental data. The authors consider the hydrodynamical behaviour and
transport processes which influence the cell
growth in bubble columns. Experimental
results and relations are presented needed for
the calculation of the most important
parameters. (112 references)

R.T. Acton, J.D. Lynn: *Description and Opera-tion of a Large-Scale Mammalian Cell Suspen-sion Culture Facility*
The parameters are described for the controlled
growth of various murine lymphoblastoid cell
lines. The facility can routinely generate in
excess of 1012 cells/week utilizing semi-conti-
nuous culture conditions. (30 references)

Sh. Aiba, M. Okabe: *A Complementary Approach to Scale-Up, Simulation and Optimi-zation of Microbial Processes*
A synthetic approach to bioreaction process
design is described, in which the most economical design is sought, taking into account inter-
action among various items of equipment and
operations of the process. (20 references)

L. Kjaergaard: *The Redox Potential: Its Use and Control in Biotechnology*
The review describes the theory for and the
measurement of the redox potential and the use
and control of redox potential in biotechnology.
(53 references)

Springer-Verlag
Berlin
Heidelberg
New York

Structure and Bonding

Editors: J.D. Dunitz,
P. Hemmerich, J.A. Ibers,
C.K. Jørgensen, J.B. Neilands,
D. Reinen, R.J.P. Williams

Springer-Verlag
Berlin
Heidelberg
New York